Variation of mechanical properties in oak boards and its effect on glued laminated timber

Variation of mechanical properties in oak boards and its effect on glued laminated timber

Application to a stochastic finite element glulam strength model

Von der Fakultät für Bau- und Umweltingenieurwissenschaften der Universität Stuttgart zur Erlangung der Würde eines Doktors der Ingenieurwissenschaften (Dr.-Ing.) genehmigte Abhandlung

Vorgelegt von

Cristóbal Tapia Camú

aus Berlin

Hauptberichter: Prof. Dr.-Ing. Harald Garrecht

Mitberichter: Prof. Dr. Frank Lam

Tag der mündlichen Prüfung: 25.01.2022

Materialprüfungsanstalt Universität Stuttgart

Institut für Werkstoffe im Bauwesen Universität Stuttgart

2022

Bibliografische Information der Deutschen Nationalbibliothek
Die Deutsche Nationalbibliothek verzeichnet diese Publikation in der
Deutschen Nationalbibliografie; detaillierte bibliographische Daten sind im Internet über
http://dnb.d-nb.de abrufbar.
1. Aufl. - Göttingen: Cuvillier, 2022
Zugl.: Stuttgart, Univ., Diss., 2022

Nonnenstieg 8, 37075 Göttingen
Telefon: 0551-54724-0
Telefax: 0551-54724-21
www.cuvillier.de

1. Auflage, 2022
Gedruckt auf umweltfreundlichem, säurefreiem Papier aus nachhaltiger Forstwirtschaft.

ISBN 978-3-7369-7588-0
eISBN 978-3-7369-6588-1

Contents

Zusammenfassung

Der nachwachsende Rohstoff Holz und hieraus hergestellte holzbasierte Bauprodukte (HBP) werden weithin als tragende Säule des nachhaltigen Bauens anerkannt. Aufgrund einer stark wachsenden Nachfrage und der mechanischen Vorteile gewinnt die bisher im Vergleich zu Nadelhölzern weniger genutzte Ressource *Laubholz* für HBPs zunehmend an Bedeutung. Gegenstand der durchgeführten Untersuchungen ist die Spezies Weißeiche (Quercus robur, Q. petraea), die nach der Buche (Fagus sylvatica) den zweitgrößten Laubholzbestand in Europa darstellt. Diese Arbeit adressiert die Erfordernis eines verbesserten Verständnisses und einer besseren Modellierung der Variabilität von Steifigkeit und Festigkeit entlang und zwischen Brettern sowie die daraus resultierenden Auswirkungen auf den Größeneffekt von Eichen-Brettschichtholz (BSH).

Für die Untersuchungen zur Variation der mechanischen Eigenschaften entlang der Hauptachse der Bretter wurde ein Satz von 53 Eichen-Brettern (Quercus robur) verwendet. Bei jedem Brett wurde die Position und Geometrie der Äste detailliert erfasst; diese Informationen wurden sodann zur digitalen Rekonstruktion der Äste verwendet. Die Elastizitätsmodul (E-Modul) parallel zur Faser wurde bei Zugbeanspruchung an 15 aneinander anschließenden 100 mm langen Brettsegmenten gemessen. Die Bretter wurden bis zum Zugversagen geprüft Die Bruchstücke wurden sodann, wenn möglich, in weiteren Zugprüfungen getestet. Auf diese Weise wurden mehrere Zugfestigkeitswerte pro Brett ermittelt.

Auf der Grundlage der E-Modul-Ergebnisse wurde ein autoregressives Model erster Ordnung [AR(1)] für die Simulation lokaler E-Modul-Profile entlang eines Brettes entwickelt. Das Model berücksichtigt die Nicht-Stationarität der E-Modul-Profile mittels einer zweistufigen Methode. Zunächst wird ein Gaußscher AR-Prozess durchgeführt, der dann in einer normalisierten E-Modul-Verteilung abgebildet wird. In einem zweiten Schritt wird das Ergebnis so skaliert, dass es einem vorgegebenen globalen E-Modul entspricht. Die Zugfestigkeitswerte wurden mittels Ereigniszeitanalyse (*Survival Analysis*) analysiert, wobei unterschiedliche parametrische und regressive statistische Modelle angepasst wurden. Die modellierten Zugfestigkeiten wurden sodann mittels eines Kreuz-Korrelationskoeffizienten mit den lokalen E-Modul-Werten gekoppelt, womit ein modifiziertes Vektor-Autoregressives Model (VAR) für den lokalen E-Modul und die Zugfestigkeit erhalten wurde. Nummerische Simulationen mit den angepassten Zugfestigkeitsmodellen ergaben einen vergleichsweise hohen Größen-, d.h. Längeneffekt, der durch einen Größenexponenten von rd. 0.23 für das 5%-Quantil charakterisiert wird.

Es wurde ein stochastisches Finite-Element-Model zur Analyse von BSH-Trägern entwickelt. Das Model berücksichtigt die mittels eines VAR-Models generierte lokale Variation der mechanischen Eigenschaften entlang jeder einzelnen Lamelle sowie die stochastische Verteilung der Keilzinkenverbindungen aneinander anschließende Bretter. Zur Berücksichtigung der Schädigungsentwicklung in den Holz- und Keilzinkenelementen, wurde ein einfacher Energie-basierter bruchmechanischer Ansatz verwendet. Das Model wurde an Versuchen mit Eichen-BSH-Trägern mit drei unterschiedlichen Querschnittsgrößen, die an der MPA Universität Stuttgart durchgeführt wurden, kalibriert. Nachfolgend wurde das Modell auf einen zweiten bei FCBA, Frankreich, geprüften Datensatz von Eichen-BSH angewandt. Die durch das Model vorhergesagten Ergebnisse stimmen gut mit den Experimenten überein. Insbesondere wird hierbei der Größeneffekt der Trägerhöhe richtig dargestellt. Der Einfluss der Materialmodelle für Holz und Keilzinkungen wurde parametrisch untersucht. Es wurde gezeigt, dass die untere Verteilungsregion der lokalen Zugfestigkeitsverteilung die Biegefestigkeitsverteilung des BSH am meisten beeinflusst. Dies ist vorteilhaft, da die untere Verteilungsregion mittels Ereigniszeitanalyse vergleichsweise präzise abgeschätzt werden kann, während die obere Verteilungsregion weitere Annahmen erfordert.

Der Autor hofft, dass die vorliegende Arbeit dazu beiträgt, die Diskussion zur Modellierung von tragenden Bauprodukten aus Laubholz zu befördern.

Abstract

Scope

The renewable material wood and hereof derived structural engineered wood products (EWPs) is widely acknowledged as being the major pillar of sustainable building construction. Due to the strongly increasing demand and technical assets the wood resource *hardwoods*, previously less used as compared to softwoods, is gaining a high momentum for EWPs. Here, the species white oak (Quercus robur, petraea) representing beside beech (Fagus sylvatica) the largest hardwood stocks in Europe is investigated. This work addresses the need of improved understanding and modeling of the variability of stiffness and strength along and between boards and the resulting impact on the size-effect of glued laminated timber (GLT) made of oak.

Experimental

A set of 53 oak boards (Quercus robur) was used to study the variation of mechanical properties along the board's main axis. For each board, detailed information regarding size and position of knots was obtained, which was then used to digitally reproduce the geometry of the knots. The modulus of elasticity (MOE) parallel to the fiber was measured in tension along each board in 15 consecutive segments of 100 mm in length. The boards were tested in tension until failure and the remnants were then tested in secondary tension tests, when possible. Thus, multiple values for tensile strength were obtained per board.

Based on the MOE results, a first order autoregressive [AR(1)] model for the simulation of local MOE profiles within board was developed. The model considers the non-stationarity of the MOE profiles by means of a two step method. Firstly, a Gaussian AR process is conducted and then mapped to the *normalized MOE* distribution. In a second step, the result in scaled to fit a specified global MOE value. The tensile strength data was analyzed by means of survival analysis, where different parametric and regression type statistical models were fitted. The tensile strength models were coupled to the localized MOE AR(1) model by means of a cross-correlation coefficient, thus obtaining a modified vector autoregressive (VAR) model for the local MOE and tensile strength along board. Numerical simulations with the fitted tensile strength models predicted a relatively high size effect, i.e. length effect, characterized by a size-effect exponent of around 0.23 at the 5 %-quantile level.

Stochastic FE model for GLT

A stochastic finite element model for the analysis of GLT beams was developed. The model considers the local variation of mechanical properties within each lamination, simulated by the derived VAR model, as well as the stochastic distribution of finger-joints connecting adjacent boards. A simple energy-based failure mechanism is considered for the evolution of tensile damage in wood and finger-joint elements. The model was calibrated with experiments of oak GLT beams of three different cross-sections tested at the MPA, University of Stuttgart, and then applied to simulate a second database of oak GLT beams tested at FCBA, France. The results obtained with the model are in good agreement with the experiments. In particular, the *size effect* of beam depth is correctly represented. The influence of the used material models for wood and finger-joints was analyzed parametrically. It is shown that the lower tail of the local tensile strength distribution, which can be estimated rather accurately by survival analysis dominates the GLT bending strength.This is fortunate, as the lower tails can be estimated by means of survival analysis in a rather accurate manner, while the upper tails require further assumptions.

Vision

The author hopes that the presented work contributes to stimulate the discussion on modelling of structural timber elements made of hardwoods.

ACKNOWLEDGEMENTS

This is the result of my research endeavors of the past years at the Division of Timber Constructions of the Materials Testing Institute, University of Stuttgart. Although the path leading to this point was not always smooth, it was certainly a very rewarding experience from both a technical and human perspectives. The first time I came to Stuttgart as a student in 2010 I was supposed to stay only for one semester—which I extended for another two—and I never thought that I would later come back and stay here for so long. I started working on the subject of this thesis in 2015 without knowing that it would eventually become my Ph.D. thesis, nor that I would enjoy so much working on it. This only shows how difficult it is to know the effects of our decisions and actions in the long term. Even more so in science! ('Will my research have any meaningful impact?' 'Am I in the right track to solve this problem?' ...<*Insert your researcher existential doubt here*>.) It is easy to feel overwhelmed or constantly considering to change plans—do I even have a plan? I should say, in my defense, that I do have some vague ideas that help me point the general direction I want to follow, however, the main reason for having survived this long in the scientific world is most certainly owed to the people surrounding me. During my time working on this thesis I counted with the invaluable support of so many people that (maybe even unaware of it) helped me succeeding in this task. To all those people I am very grateful.

Firstly, I thank my advisor Dr. rer. nat. Simon Aicher for his constant support throughout all these years. As an internationally highly regarded researcher in the field of timber constructions his disposition to share his experience and openly discuss different matters was fundamental to develop the ideas presented here, and for this I am deeply grateful.

I want to express my gratitude to Prof. Dr. Frank Lam (University of British Columbia) for having accepted being part of the Examination Committee and for sharing his ideas for future research topics related to my work.

I thank Prof. Dr. Garrecht not only for accepting being the principal Examiner in my defense, but also for showing interest in my research and future work.

I am indebted to Prof. Dr. Anders Olsson and Dr. Jan Oscarsson from Linnæus University for the inspiring discussions on variation of mechanical properties in boards, and to Wolfgang Klöck (MPA, University of Stuttgart) for sharing his knowledge on statistics and survival analysis. Also to Dr. Guillaume Legrand and Dr. Carole Faye (FCBA, Bordeaux) for allowing me to use their experimental results of glued laminated timber beams made of oak.

Thanks to Matthias Wessel and Ulrich Wachter-Sigel from the technical staff for their helpful disposition and to the former working students Manuel Magdanz and I-Ching Lee for their help with measurements and documentation. I am grateful to all my colleagues at the Division of Timber Constructions at the MPA, University of Stuttgart, for their constant support, fun coffee (*tea*) breaks and always friendly working environment. In particular I would like to thank Dr. Gerhard Dill-Langer, Jan Hamming, Maximilian Henning, Martin Hentschke, Nikolai Ruckteschell, Jürgen Hezel, Dr. Nicola Zisi, Roland Falk and Dominique Sghair, as well as former colleagues Gordian Stapf, Cyrill Stritzke and Maren Hirsch, for their help in one way or another during this years-long journey.

I thank the German foundation FNR, Fachagentur Nachwachsende Rohstoffe e.V., for their support through the contract 2200414, in conjunction with the European ERA-WoodWisdom project "European hardwoods for the building sector (EU Hardwoods)".

Finally, not a single page of this thesis would have been possible without the unconditional support from my family throughout my entire life. Thanks for everything!

Stuttgart, March 2022 — Cristóbal Tapia Camú

"I do not know why a certain event occurs; I think that I cannot know it; so I do not try to know it and I talk about chance."

— Lev Tolstoy, *War and peace*

INTRODUCTION

1.1 European forests in the 21st century

Forests in Europe are currently in the midst of a significant change in their tree-species distribution (Cheaib et al., 2012; Delzon et al., 2013). The increase in global temperatures in the last century and the changes in local weather conditions deriving from it (e.g. rain patterns, drought periods and winds) have already started showing some effects (Lindbladh et al., 2000). The consequences that the current climate change will ultimately bring to the European forests are largely unknown; the amount of variables that play a relevant role is considerable (Lindner et al., 2014)—alone the prediction of future temperatures depends on highly sensible parameters like estimated greenhouse gases emissions. In spite of these uncertainties, enough scientific evidence points to major changes in the forest, characterized by steady geographical shifts, as well as expansions and contractions of the habitable regions of species (Meier et al., 2011). This dynamic changes the distribution of species in the forests, and consequently affects the silviculural activity in a direct manner.

Under the current and projected conditions, the habitable ranges of deciduous species (hardwoods) are expected to shift far northern with respect to their current boundaries. Field studies by Delzon et al. (2013) have confirmed a steady colonisation of Holm oak (*Quercus ilex*) northwards from its natural range in the last century, and simulations predict a probable ongoing of this development during this century (Cheaib et al., 2012). Meanwhile, coniferous species ranges will be confined to higher altitudes (Lexer et al., 2002) and latitudes (Delzon et al., 2013), where more suited climatic conditions are to be expected.

Although important, global warming is not the only driving force reshaping the forests. The silvicultural and forestry activity of the last century is largely accountable for the observed increase in disturbance impact in the European forests, mainly due to restructurations of the distribution of species (Seidl et al., 2011). The favoring of fast-growing species, e.g. Norway spruce (*Picea abies*) or Scots pine (*Pinus sylvestris*), which ensure small rotation periods, led to a displacement of native species from their natural environments and gave place to large monocultural forests all over Europe (Felton et al., 2010). This presents its own set of associated problems, e.g. an increased probability of large, concentrated die-backs of species due to a higher vulnerability to both biotic (pest outbreaks) and abiotic (winds, droughts) factors (Felton et al., 2010). The predicted climate change can only amplify these problems (see e.g. Lexer et al., 2002).

The understanding of the consequences of these two driving forces marked a turning point in the management of European forests regarding its economical, environmental and social uses (CEC, 2007). It became evident that forests need to adapt to the future climatic conditions, yet the natural process to achieve this—through natural dispersion of seeds—would not be enough to keep up with the predicted speed of climate changes (Meier et al., 2011; Delzon et al., 2013). Thus, the adaptation process needs to be assisted. Different guidelines exist to transit to more climate resilient forests in Europe (see e.g. LFBW, 2014). In general, mixed-species stands are now strongly encouraged, with the objective of reducing the vulnerability to diverse risks and promote biodiversity (CEC, 2007), while at the same time maintaining an economical competitiveness with respect to e.g. spruce monocultures (Agestam et al., 2006).

The interaction of climate change and these new forestry strategies will lead to a much higher share of hardwoods in the European growing stock, including oaks, beech and birch species, among others (Cheaib et al., 2012). In fact, in the last decades European forests have seen a steady increase in their share of deciduous species, which can be quantitatively observed i.a. in national forest inventories of different countries. For example, between the first and second German National Forest Inventory (1987–2002) Norway spruce has lost about 7 % of its total area to deciduous species in south-west Germany (Lindner et al., 2014). For the timber building industry, up until now clearly dominated by softwood species, this development presents many challenges and a need for adaption.

1.2 Hardwoods in the building industry

The described gradual transition to a higher proportion of deciduous species in the forests has motivated a small, yet steadily increasing proportion of the European timber industry to incorporate more hardwood products into their catalogs. This trend is further reinforced by additional factors, such as (i) the fact that often the mechanical properties of hardwoods are significantly superior to those from softwoods (Aicher and Stapf, 2014), and (ii) that from an aesthetical point of view many hardwoods are typically considered to be more visually appealing. In order to harvest the full potential of hardwoods, improvements at both technical and regulatory levels are required.

Consequently, the research output dealing with different aspects of the value chain of hardwoods has shown a steady increase in recent times. The topics are multiple and include e.g. in-depth studies of material availability, improved classification methods, determination of mechanical properties and development of engineered products such as glued laminated timber (GLT) and, to a minor degree, cross-laminated timber (CLT), too. In this context, the project "European hardwoods for the building sector" (EU Hardwoods, 2017)—where the origins of the present work can be found—was tasked with analyzing these topics in a holistic manner. The focus was then placed on a subset of the most relevant European hardwood species, consisting on beech (*F. sylvatica*), oak (*Q. robur, Q. petraea*), chestnut (*C. sativa*) and ash (*F. excelsior*). There, an assessment of the suitability of these species for structural applications in the form of engineered products was made.

The past decade has seen an increase in the efforts of bringing hardwood engineered structural timber products to the market. In Germany and Austria, special attention has been given to beech, being this the most abundant deciduous species in these countries, where both GLT and CLT have been an important research subject (see e.g. Frese, 2006a; Aicher et al., 2016; Ehrhart, 2020). For the case of oak wood, being the primary hardwood species in France and second most important in Germany, an increased interest has been observed, too (Aicher and Stapf, 2014; Faydi et al., 2017). Although oak was initially limited to GLT members with rather small cross-sections—mainly used as post and beam window façade elements—it quickly evolved to include larger cross-sections for structural applications (Aicher et al., 2014).

Further research has shown that an efficient way to incorporate hardwoods into structural elements is by means of so-called *hybrid elements*, where both softwoods and hardwoods—currently mostly beech—are used together, taking advantage of the superior mechanical properties of hardwoods where it is needed. For GLT beams this concept is applied by replacing the material of the outer laminations with hardwoods of high tensile strength (e.g. Blaß and Frese, 2006). In CLT plates, more interestingly, the middle cross-layers, where the failure mechanism is dominated by rolling shear, can be substituted by e.g. beech material that would normally be regarded of low quality, yet presents a much higher resistance to rolling shear (e.g. Aicher et al., 2016).

Although these concepts present a high relevance for the future of engineered timber elements, the landscape of hardwood products in the coming years will probably continue to be dominated by the more simple single-species products, especially in the form of glued laminated timber.

1.3 Glued laminated timber

Glued laminated timber (GLT) is currently one of the most used wood-derived engineered products in the world when considering beam-like applications. It is produced by connecting a series of boards lengthwise by means of finger-joints to form a so-called *endless lamella*. This lamella is cut at regular intervals, defined by the length of the beam to be produced. The obtained parts are stacked on top of each other, applying glue between each layer (see Fig. 1.1). A constant pressure is then applied normal to the wide lamella faces throughout the so-called *minimum pressing time*. This ensures a sufficient bond strength and enables a further curing of the adhesive without the need of additional pressure, including careful transportation.

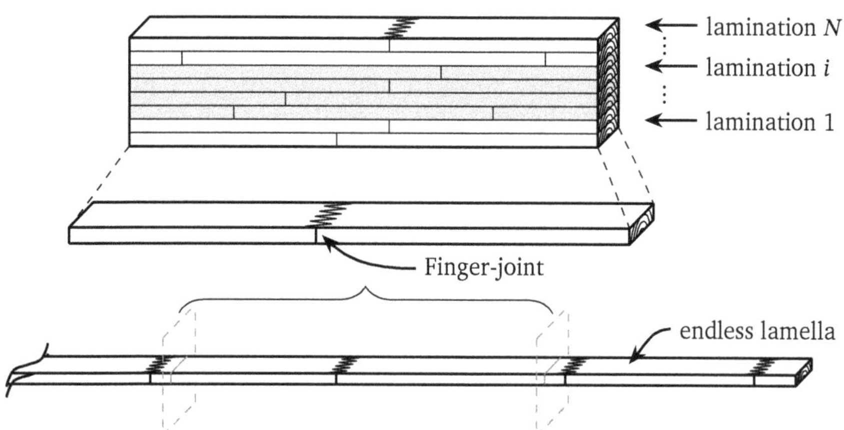

Figure 1.1. Description of the engineered timber product glued laminated timber

Compared to solid timber, the build-up of GLT allows for higher flexibility in terms of possible geometries and maximum spans, the latter being commonly limited to about 25 m due to transportation constraints. A further advantage is commonly referred to as *homogenization* of the material, which is associated with a reduction in the variability of the mechanical properties; e.g. low stiffness of one board is compensated by possible higher stiffness of adjacent laminations, thus the variation of stiffness at the beam level is reduced. A further relevant concept is known as the *lamination effect*. This corresponds to the positive mechanical effect obtained from vertically stacking lamellas, and originates mainly from two aspects: (i) the restraining of lateral deformations of individual laminations (see Fig. 1.2) due to local defects—thus preventing what would otherwise mean a reduction of tensile resistance (Foschi and Barrett, 1980)—and (ii) the redundant nature of a parallel system, where local failures in one lamination can be compensated to some degree by the adjacent laminations. The direct consequence of the lamination effect is the fact, that the observed bending strength of GLT beams is higher than that of the individual laminations.

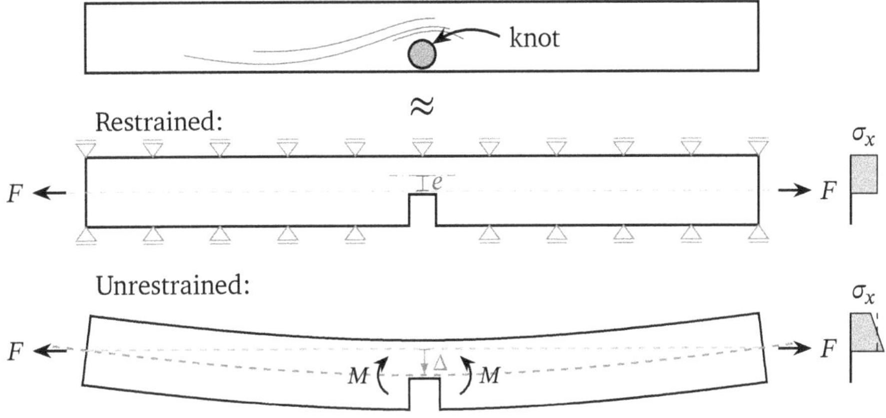

Figure 1.2. Lateral deformation of board in tension due to the presence of a knot on an edge (see also Foschi and Barrett (1980))

The degree of utilization of the material can be further increased by using so-called *composite build-ups*, characterized by the allocation of laminations of different strength grades—from the same species—throughout the cross-section. This is illustrated by the different colors of the laminations in Fig. 1.1. Specifically, boards of higher grades are placed in the outer zones (subjected to higher bending stresses), whilst lower grades can be used in the central part. The presence of lower grade boards in the central region of the cross-section has no negative impact in the shear resistance of the beam, as the shear strength remains practically the same for the different board grades. In this manner, an under-utilization of the higher quality material is reduced, and the usage of the material is optimized. The mechanical properties of such composite, as well as for homogeneous build-ups are specified in EN 14080 (2013).

1.3.1 Characterization of mechanical properties of glulam according to EN 14080

For the European market the build-up, strength features and minimum production specifications of GLT products is regulated by EN 14080 (2013). This standard defines minimum requirements on finger-joints and boards for the production of GLT with specific mechanical characteristics, defined as *GL* classes. GL classes are defined by their characteristic, i.e. 5 % quantile bending strength, $f_{m,g,k}$, meaning e.g. that a GLT beam of the class "GL24" presents a $f_{m,g,k}$ value of 24 N/mm^2. Homogeneous build-ups are marked by appending the letter "h" to the GLT class (e.g. GL24h), whilst composite build-ups, consisting of up to three lamination strength-grades, are denoted by the letter "c" (e.g. GL24c). The characteristic value attributed to each class is defined for a cross-sectional depth of 600 mm.

In order for the producer to declare the appropriate GL class, EN 14080 (2013) defines three possibilities:

1. The first option comprises the classification of the build-ups and their lamination properties according to tabulated values.
2. The second option is based on the use of a simulation-based, empirically adjusted equation relating the mechanical properties of finger-joints and boards to obtain the characteristic bending strength of the GLT.
3. The third alternative consists on the experimental determination of the GLT characteristic properties.

It is rather clear that the producers would strongly prefer either of the first two alternatives, as the realization of experimental tests is rather expensive. The equation mentioned in the second option relates the strength of both, finger-joints and boards with $f_{m,g,k}$ as

$$f_{m,g,k} = -2.2 + 2.5 \cdot f_{t,0,k}^{0.75} + 1.5 \cdot \left(\frac{f_{m,j,k}}{1.4} - f_{t,0,k} + 6 \right)^{0.65} , \qquad (1.1)$$

where $f_{t,0,k}$ and $f_{m,j,k}$ are the characteristic values for tensile strength of the boards and flat-wise bending strength of the finger-joints, respectively. In addition, the following restriction is specified for the finger-joints:

$$1.4 \cdot f_{t,0,k} \leq f_{m,j,k} \leq 1.4 \cdot f_{t,0,k} + 12 \quad . \tag{1.2}$$

Since the relation $f_{m,j,k} = f_{t,j,k} \cdot 1.4$ is implicitly assumed, the condition (1.2) means that the characteristic tensile strength of the finger-joint, $f_{t,j,k}$, must be higher than the characteristic tensile strength of the boards. In essence, the first alternative is nothing more than the application of Eq. (1.1) to a set of specific combinations of finger-joint and board strength values, and build-ups.

Although CE-marking of GLT beams according to EN 14081-1 (2016) is only possible for softwood GLT, the foreword of the standard mentions a *principle of applicability* to hardwood GLT, too. Regarding hardwood GLT, however, concerns on the validity of Eq. (1.1) have been raised, especially when considering the restriction of Eq. (1.2). This is owed to the empirical fact, that hardwoods, reaching in general much higher tensile strength values than the usual softwoods, often present problems at achieving the required (rather high) strength values for the finger-joints (Aicher and Stapf, 2014). As a consequence, for such cases, finger-joints represent the weakest region of the GLT, altering the statistical characteristics of the measured bending strength values. However, this does not mean that such GLT is not apt for structural use.

1.3.2 Towards standardization of hardwood GLT

Since the requirements specified in EN 14080 (2013) for softwoods cannot always be fulfilled by hardwoods (see above), alternative procedures have to be used to certify hardwood GLT. Currently the only available options consist in obtaining either a National Technical Approval or a European Technical Assessment (ETA) on the basis of a European Assessment Document (EAD)[1] (EU, 2014). Recently, a few producers have used these options to certify their production of GLT, e.g. for oak (Z-9.1-704, 2012; Z-9.1-821, 2013; ETA-13/0642, 2013) and chestnut (ETA-13/0646, 2013). However, the relative high costs associated with such a certification process, mainly due to a large number of required experimental full-scale tests, pose a clear barrier for many glulam producers. Engineered products from hardwoods, with all their structural advantages, can only become relevant if

[1]Up to the year 2015, i.e. to the end of the European Building Products directive (EEC, 1988), the basis of an ETA, then called European Technical Approval, was the Common Understanding Approval Procedure (CUAP)

the proper regulations are adapted for their use, which points to the need for a standardization.

Therefore, the European Committee for Standardization has mandated the Technical Committee 124, Working Group 3, Task Force 1, with the development of a first standard for the regulation of the production of hardwood GLT. The idea is to create a document with provisions analogous to EN 14080 (2013) but considering the specifics of the different hardwood species and the limitations of the current adhesives. This is easier said than done, as the differences within hardwood species is known to be far larger than in softwood species. These differences span from the cellular level to the observed growth-bound defects, such as knots and fiber deviations, and have a direct influence in how the material behaves. Thus GLT made of hardwood might require slightly adapted models for each species. Although this is a difficult task, inspiration can be found in the process that lead to the standardization of softwood GLT.

1.3.3 The example of softwoods: strength models

The research pathway that led to the empirical model between strength properties of boards and finger-joints, and the bending strength of the glulam beam defined in EN 14080 (2013) has a long trail that dates back to the early 1980's. The first reported glulam strength model that used computer simulations to consider the stochastic variability of the material was introduced by Foschi and Barrett (1980). This model divided each lamination into cells of a determined length and assigned a MOE and strength value to each of them, which were correlated to generate knot and density values by means of a function. Although the influence of finger-joints was not considered, the stochastic essence of the variation of mechanical properties along the boards defined the nature of the models to come. Of high relevance for the European standardization was the model presented by Ehlbeck et al. (1985), the so-called "Karlsruher Rechenmodel", based on an extensive study of the variation of the mechanical properties of boards and finger-joints of Norway spruce (*Picea abies*). There, several correlations between strength, modulus of elasticity (MOE) and grading criteria were established, allowing the simulations to better represent the distribution of properties in glulam beams. After a long period of calibration, the numerical model was finally implemented as Eq. (1.1), relating GLT bending strength with strength values of both, finger-joints and boards in Section 5.1.5 of EN 14080 (2013).

Further models have been developed since then, i.a. by Hernandez et al. (1992) and more recently by Fink (2014), both calibrated with softwood species. Blaß et al. (2005) presented the first adaptation of the "Karlsruher Rechenmodel" to a hardwood species (beech) and later applied it to a hardwood-softwood hybrid build-up (Blaß and Frese, 2006), manifesting the utility and versatility of a stochastic FE-based strength model. In all these models the correct characterization of the material properties of the boards and finger-joints plays an essential role to correctly simulate the behavior of the studied GLT material. Substantial research has been dedicated to this aspect, i.a. Showalter et al. (1987), Taylor and Bender (1991), Lam and Varoglu (1991a), Isaksson (1999), and Fink (2014), gathering large empirical databases and developing different methods to simulate the observed variation of properties within boards. The present work is inspired by these investigations and presents a new approach to analyzing and modelling the variation of properties within board, using oak as an example material. The simulated mechanical properties can then be applied to a softening-based glulam strength model in order to study the bending behavior of the simulated GLT.

1.4 Objectives

This thesis presents the implementation of a new glulam strength model for hardwoods. For this, experimental results on boards and finger-joints are described and analyzed in order to obtain correlations between different properties at a global level. A method to generate the needed properties considering the correlations between them and their statistical distributions is presented. Investigations on the variation of properties within single boards are analyzed as well, and a method to simulate the observed variation is developed.

A finite element model with fracture mechanics capabilities is implemented using the software Abaqus. The input data is generated using a combination of the global and local material models mentioned above. Characteristic bending strengths are obtained for different build-ups by the application of the Monte Carlo method, and the results are compared to experimental investigations, for which the distributions of material properties are known. Finally, the sensitivity of the model to different parameters is assessed by means of a parametric analysis.

To summarize, the objectives of this thesis are:

- Analyze the distribution of defects in oak boards,
- Analyze the variation of density, modulus of elasticity (MOE) and tensile strength within oak boards,
- Characterize the autocorrelation parameters of MOE along board,

- Characterize the variation of tensile strength within board and its dependence with localized MOE values,
- Develop a simulation model capable of generating MOE and tensile strength profiles, representing oak boards with specific characteristics, and
- Develop a finite element simulation model for the prediction of bending strength, considering softening behavior.

1.5 Scope

The present study investigates the mechanical properties of boards and glulam beams made of oak (*Quercus robur, Quercus petraea*). In this sense, the parameters obtained for the material model are limited to the studied species. However, the material and finite element models presented in this work should be generally applicable to other wood species—assuming that the required calibration has been performed.

1.6 Outline and overview

The structure of the thesis is designed to follow the development of the strength model in a logical manner. It starts with a review of the state of the art in Chapter 2, where previous glulam strength models are reviewed and the needed statistical background is presented. Chapter 3 introduces the materials used for both the glulam testings and the study on variation of properties along boards. The distribution of knots in boards is studied in Chapter 4. The variation of properties along boards is analyzed in Chapter 5. A model for the simulation of the properties along boards is presented in Chapter 6. The finite element model for the determination of bending strength is introduced in Chapter 7, where the different components and parameters are explained. The calibration of the model with the experimental data is done in Chapter 8, where also diverse parameters of the strength model are studied in order to observe the susceptibility of the model with regard to the parameters. Final conclusions and an outlook are presented in Chapter 9.

STATE OF THE ART

2.1 General remarks

This chapter introduces the most relevant models present in the literature for the analysis and simulation of the mechanical properties for boards and GLT beams. It starts by revealing the diverse sources of variation for the mechanical properties of wood, especially for the case of hardwoods. Then, some general statistical concepts and methods are introduced, which are needed to understand some of the models described later. Existing models to simulate the variation of mechanical properties along boards are then presented. Finally, the application of these models to the stochastic simulation of GLT beams is shown, where the different strength models are explained in detail.

2.2 A look at the variation in wood: sources of uncertainty

Wood is a natural-grown material, characterized by a marked anisotropy in its mechanical properties that can be traced down to the cellular level, to the shape of the individual cells and their growing pattern. The latter is characterized by the yearly growth of concentric layers of *earlywood* and *latewood* in temperate climate zones. Wood cells present different shapes and sizes depending on their specific task withing the tree. Nevertheless, the chemical composition of their cellular walls is very similar, primarily composed of layers of cellulose strands (*microfibril*), lignin and hemicellulose. For the case of softwoods, mainly two cell types are present: (i) string-like cells called *tracheids* being many times longer than wide (about 5–7 mm in length), and (ii) *rays*, presenting a rectangular prismatic shape. These two cell types are arranged in a rather simple structure, where the tracheids are aligned in longitudinal direction (i.e. parallel to the stem axis), delivering structural stability and vertical transport, while the rays are oriented radially and provide mainly storage and synthesis of biochemicals. The tracheids amount rather consistently to about 90 % of the wood volume in all softwoods (Ross, 2010), meaning that at the microstructure level all softwoods are very alike.

Hardwoods, in contrast, present a much wider spectrum of cell types, and their arrangements and proportions related to total volume greatly vary between species. Compared to softwoods, hardwoods exhibit a higher level of specialization in the structural and conductive functions (Rao et al., 1997), which are mainly provided by the fibers and vessels, respectively. For the particular case of *Quercus robur*—a European oak representing a *ring-porous hardwood* species—the cellular diversity comprises vessels, *libriform* fibers, tracheids, fiber-tracheids and ray/axial parenchyma cells (Gričar et al., 2013). Their proportions and arrangements differ clearly for earlywood and latewood, and their ratios in each new ring change as the tree grows older. Specifically, the proportion of vessels in latewood of newer rings increases while the proportion of libriform fibers (source of strength) decreases. This leads in theory to a reduction of the mechanical properties of the younger rings (Rao et al., 1997), producing a negative strength gradient from the stem's center to the cambium, defined as the region between the last grown ring and the inner bark. Experiments on oak boards sawn from trunks of different diameters presented by Lanvin and Reuling (2012) clearly show this effect.

As the tree grows, diverse environmental and genetic factors affect the precise arrangement and chemistry of the wood cells. For example, as the years pass, the older cells in the inner rings begin transitioning from sapwood to heartwood, process whereby the cells slowly loose their protoplasts and die, generally leading to an improvement of the mechanical properties of wood. Furthermore, for pronouncedly inclined trees—due to high loads or topographical reasons—localized changes at the cellular level can develop, producing *reaction wood*. In hardwoods this occurs in the form of *tension wood*, a region characterized by the development of strong gelatinous fibers in the upper side of an inclined tree, i.e. in the bending tension zone (Kollmann and Côté, 1968). Considering the proportions of different cell types in each ring of hardwoods, they can be affected e.g. by the growth rate

(Rao et al., 1997).

The high anisotropy exhibited by wood, where the mechanical properties parallel to the main axis of the fiber cells (grain) are considerably higher than in the other two orthogonal directions (radial and tangential), produces a high sensitivity of the material to local and global perturbations of the grain. Deviations of the grain occur for different reasons, and can span a rather large region or be very localized. For example, long-standing or cyclical loads applied to a tree, e.g. by wind action, can induce deviations in the grain direction throughout the entire trunk due to an excessive torsion of the stem. The orientation of fibers can further be altered by the growing of branches—commonly known as *knots* in structural timber. These deviations have a local character and the study of its effects in the mechanical properties of timber has a high relevance, reason for which it has been extensively investigated by means of different approaches e.g. by Foschi and Barrett (1980), Ehlbeck and Colling (1987), Lam et al. (2005), Fink et al. (2011), Olsson et al. (2018), Lukacevic et al. (2019), and Wright et al. (2019). At a macroscopic level this constitutes the most evident source of variation in timber.

Within the frame of industrial processing of structural lumber for engineered wood products additional sources of variation develop, increasing the variability of mechanical properties. Examples are the chosen sawn pattern and the so-called *seasoning*, process by which the moisture content of wood is brought to an equilibrium with its surroundings—this can be done e.g. by natural or kiln-drying. Thus, the boards finally obtained for structural use are the product of multiple processes, resulting in a rather large uncertainty in the structurally relevant mechanical properties.

Diverse classification methods have proved to reduce this uncertainty at the board level. Statistical methods have been applied to different variables, effectively increasing the yield of the material. However, the variation of mechanical properties along a board will always be present and needs to be accounted for in the production of glued timber products.

The study of the variation of mechanical properties along boards implies specially designed tests and a deep statistical analysis of empirical results. This has been done mostly for softwood species i.a. by Ehlbeck et al. (1984), Foschi and Barrett (1980), and Fink (2014) and for hardwoods (beech) by Frese (2006a), which are presented below. For oak, work on the variation of material properties has been done based on laser scans by Olsson et al. (2018), however with a focus on the global behaviour by defining new indicator properties.

2.3 General statistical concepts and methods for the analysis of wood properties

2.3.1 Weibull theory of weakest link

The theory of the weakest link can be better explained with the canonical problem of studying the load capacity of a chain consisting of n links. The probability of failure of each link at a certain load x is defined by the distribution function $F(x) = P(X \leq x)$. Due to the serial nature of the system, the failure of the chain is determined by the failure of its weakest link.

With each newly added link to the chain, the probability of having a *weaker* link rises. As a consequence, the distribution for the load capacity of the chain depends on the number n of links. This simple concept constitutes the basis for the so-called *size effect* observed in the failure of (mostly brittle) solids.

Weibull (1951) derived a statistical distribution function that takes into account this issue. The main insight was to find a distribution $F(x)$ that can be represented as

$$F(x) = 1 - e^{-\varphi(x)} \quad . \tag{2.1}$$

The probability of survival (not failure) of a chain with n links can be represented as

$$S_n = 1 - P_n, \tag{2.2}$$

with P_n the probability of failure of the chain. Since the probability of survival is equal to the simultaneous survival of each link, the following equation holds:

$$1 - P_n = (1 - P)^n \quad . \tag{2.3}$$

Assuming that Eq. 2.1 holds, the following expression for P_n is obtained:

$$P_n = 1 - e^{-n\varphi(x)} \quad . \tag{2.4}$$

The most simple mathematical expression that satisfies the needed conditions is $\varphi(x) = (x - x_u)^m / x_0$, thus obtaining

$$F(x) = 1 - \exp\left[-\left(\frac{(x - x_u)}{x_0}\right)^m\right] \quad . \tag{2.5}$$

This is commonly known as the "Weibull distribution" and is not constrained to mechanical problems. As pointed out by the original author, it is a distribution of wide applicability (Weibull, 1951). This theory can be considered as a particular case of the theory of extreme value, which is introduced later in this chapter.

2.3.2 Order statistics and extreme value theory

Order statistics can be regarded as a generalization of the weakest link theory, and deals with the statistical analysis of the r-th smallest value in a sample of size n. More formally, if $X_1, X_2, \ldots, X_n$ constitute a sample drawn from a given distribution $F(x)$ and are sorted in ascending order, such that $X_{1:n} \leq \ldots \leq X_{n:n}$, then the r-th element of this sequence is defined as the r-th *order statistic* of the sample (Castillo et al., 2005). Of special interest are the first and the last elements (minimum and maximum of $X_1, \ldots, X_n$, respectively), generally referred to as the *extreme values*.

For any given parent distribution, $F(x)$, the distribution of the minimum value for a total of n elements, $F^{(n)}_{\min}(x)$, can be simply derived from $F(x)$ as (Castillo et al., 2005)

$$F^{(n)}_{\min}(x) = 1 - [1 - F(x)]^n \, . \tag{2.6}$$

As it can be noticed, this is the same expression as derived by Weibull (1951).

2.3.3 Survival analysis

Survival analysis is a field in statistics that deals with the assessment of what is commonly known as *censored data*. Traditionally, it has been used in medical statistics to analyze the survival rates of patients to different treatments and conditions—from where the name comes from. Censoring arises when the data is analyzed while some patients are still alive, and also when a patient is lost to follow-up during the course of the study for some reason (Miller, 1976). The same

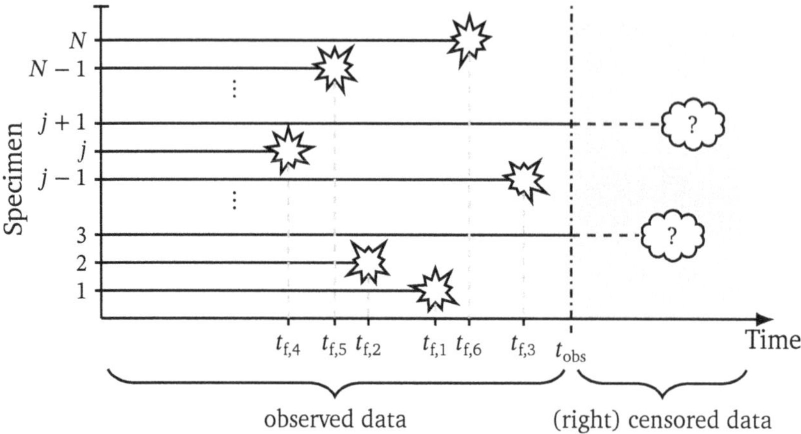

Figure 2.1. Illustration of the concept of (right) censored data in an experimental test

concepts apply in different disciplines. For example, in engineering, the time to failure (t_f) of a component in a duration-of-load test setup is the relevant observed variable. If the data is analyzed at a time t_{obs} before the end of the test, then censored data will need to be considered (see Fig. 2.1). In the field of timber structures, survival analysis has been used e.g. by Klöck (2005), where censoring occurred in shear testings of GLT beams due to some specimens failing in bending instead of shear.

In general, the problem consists on estimating the set of parameters θ of the candidate distribution function, $F(x)$, that describes the studied variable. For this, maximum likelihood estimation (MLE) is used, where the usual likelihood function ($\mathcal{L}(\theta)$) is modified to consider the censored data (Odell et al., 1992) according to

$$\mathcal{L}(\theta) = \prod_{i=1}^{n} f(x_i)^{\delta_i} \cdot (1 - F(x_i))^{(1-\delta_i)}, \tag{2.7}$$

where $f(x)$ is the probability density function (PDF) and δ_i is an indicator parameter, i.e. a binary variable, signalizing whether the data was observed or not (censored):

$$\delta_i = \begin{cases} 1 & \text{, if } x_i \text{ is an observed data point} \\ 0 & \text{, if } x_i \text{ is a censored data point.} \end{cases} \tag{2.8}$$

The main characteristic of this likelihood function is the use of the survival function $S(x) = [1 - F(x)]$ for the censored data instead of the PDF. Normally, it is easier to work with the logarithm of the likelihood function, ($\ell(\theta)$), which in this case is represented by

$$\ell(\theta) = \log \mathcal{L} = \sum_{i=1}^{n} \delta_i \cdot \log(f(x_i)) + (1 - \delta_i) \cdot \log(1 - F(x_i)). \tag{2.9}$$

Maximizing this function yields the maximum likelihood estimates, $\hat{\theta}$, for the tested distributions.

2.4 Probabilistic methods for the simulation of correlated variables

In the context of stochastic simulations, e.g. by applying the Monte-Carlo method, it is often necessary to preserve observed correlations between variables in order to better represent possible random states of a given system. The literature presents diverse methods to this end, which are applicable the simulation of variables belonging to a wide variety of statistical distributions. This section introduces several generally applicable methods, relevant for the understanding of the models described later in this chapter. They are further relevant for the following chapters, which deal with the variability of mechanical properties of boards.

2.4.1 Simulation of correlated normally distributed variables

The simulation of correlated normally distributed variables is needed in stochastic analyses when the interaction of different variables needs to be considered. One of the most commonly applied methods is based on the fact that any positive definite matrix $\mathbf{\Sigma}$ of dimensions $n \times n$ can be decomposed in a matrix $\mathbf{C}$, such that $\mathbf{C}\mathbf{C}^{\mathrm{T}} = \mathbf{\Sigma}$ (Tong, 1990). The matrix $\mathbf{C}$ can be easily found applying the Cholseky decomposition. Thus, if $\underline{\mu}$ is the vector of length n containing the mean values of n different normal distributions, $\mathbf{\Sigma}$ is the covariance matrix, and $\mathbf{Z}$ is a multivariate Standard Normal distribution, $\mathcal{N}_n(\mathbf{0}, \mathbf{I_n})$, then the matrix $\mathbf{X}$ resulting from applying the following equation

$$\mathbf{X} = \mathbf{C} \cdot \mathbf{Z} + \underline{\mu} \tag{2.10}$$

correspond to an $\mathcal{N}\left(\underline{\mu}, \Sigma\right)$ distribution.

The following algorithm—slightly modified from (Tong, 1990)—is then used to generate the correlated random variates $\underline{X_1}, \dots, \underline{X_N}$:

1. Compute the matrix **C** by applying Cholseky decomposition;
2. Generate $\mathcal{N}(0,1)$ distributed random variates for each variable and place them in the rows of a matrix **Z**;
3. Generate correlated random variates, **X**, by applying Eq. (2.10).

The resulting matrix **X** contains the correlated random variates of each variable in each row.

2.4.2 Simulation of correlated non-normally distributed variables

In many situations the variables that need to be simulated do not follow normal distribution. For such cases, the previous method can still be applied, however, with an additional step at the end, consisting in *mapping* the normal random variates into the needed distributions. If random variates, Z, from a $\mathcal{N}(0,1)$ are generated (e.g. by the previous method), then the mapping process can be concisely described by the following equation:

$$Y = F^{-1}\left[\Phi(Z)\right] \quad , \tag{2.11}$$

where $F^{-1}(\cdot)$ is the inverse CDF, or percent point function of the needed distribution, and $\Phi(\cdot)$ is the CDF of the standard normal distribution, $\mathcal{N}(0,1)$. This process is illustrated in Fig. 2.2.

This concept is sometimes referred to as *Gaussianization* (Chen and Gopinath, 2000) and is a commonly used method to translate variables from one distribution into a different distribution (e.g. Taylor and Bender, 1988; Grigoriu, 1998; Deodatis and Micaletti, 2001).

2.4.3 Autoregressive models

Autoregressive models (also known as Markov processes) are widely used across different science fields to model series of data where a correlation between each

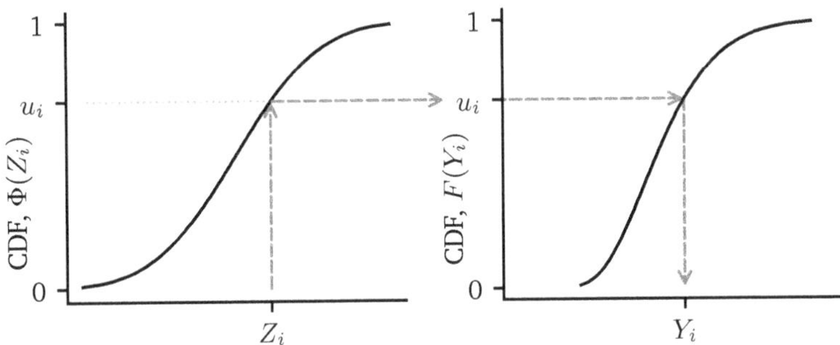

Figure 2.2. Illustration of the process of mapping a variable y_i from a normal distribution (left) to the corresponding value y_i of an arbitrary distribution (right), maintaining the same value on their respective cumulative distribution functions Φ and F.

element and the previous p data points can be established. In its most basic form an autoregressive model of order p, AR(p), is defined as:

$$x_i = \sum_{j=1}^{p} \varphi_j x_{i-j} + \varepsilon_i \quad , \tag{2.12}$$

where φ_j are the model parameters and ε_i is a white noise term, generally assumed as a zero-centered normally distributed random variable, $\mathcal{N}(0, \sigma)$. The white noise needs to be a stationary process. Each AR model possesses an autocorrelation function (ACF), which shows the correlation for the k-th lag, e.g. the k-th time or location interval. In general, the ACF, $\rho(k)$, is a function of the autocovariance function ($\gamma(k) = \mathrm{Cov}(x_{i-k}, x_i)$), as

$$\rho(k) = \frac{\gamma(k)}{\gamma(0)}. \tag{2.13}$$

For the particular case of an AR(1), the ACF is defined as:

$$\rho(k) = \varphi_1^k \quad , \tag{2.14}$$

where φ_1 is the model parameter and k is the k-th lag. This describes a geometrical progression, which is characteristic for AR(1) models.

The ACF serves as a theoretical reference point when fitting a dataset to an AR model, as it can be compared to the serial correlations computed directly from the studied data. This is known as the *sample autocorrelation function* (SACF), $\widehat{\rho_k}$, and is defined as (Brockwell and Davis, 2002):

$$\hat{\rho}(k) = \frac{\hat{\gamma}(k)}{\hat{\gamma}(0)} \quad , \tag{2.15}$$

with the sample autocovariance function

$$\hat{\gamma}(k) = \frac{1}{N} \sum_{i=1}^{N-k} (x_{i-k} - \bar{x})(x_i - \bar{x}) \quad , \tag{2.16}$$

where $\bar{x}$ is the mean of the sample and N is the total number of observations. Since the objective is to compute the SACF that will later be used to fit an AR model, it is necessary to make sure that the data describes a stationary process. If this is not the case, a common solution is to apply a *transformation* to the data, which produces the needed stationarity. A typical case of this is, for example, the consideration of seasonality (Brockwell and Davis, 2002).

2.5 Models for the simulation of material properties along boards

Timber boards, being an organic-grown material, present large variability in their material properties, both between different boards, as within a single board. Material properties such as modulus of elasticity and density vary continuously throughout the volume of the board, being influenced i.a. by fiber orientation, knots and year rings, producing a complex three-dimensional scalar field. However, for most applications, a board can be considered as a one-dimensional element, where a given property along the board's main axis (length direction) correspond to the aggregation of the property over the cross-sectional plane (width and thickness direction) at each position over the length. This is schematically illustrated by the solid line of the upper diagram of Fig. 2.3 for modulus of elasticity, $E_{t,0}$.

For the tensile strength, the reduction from three dimensions to one dimension cannot be achieved in the same manner. For in this case the load redistribution between wood fibers within a given volume has to be considered, too (see e.g. Dill-Langer et al., 2003). Thus, at most, the material strength reduced to one dimension can be represented by a step-function (semi-continuous variation), as

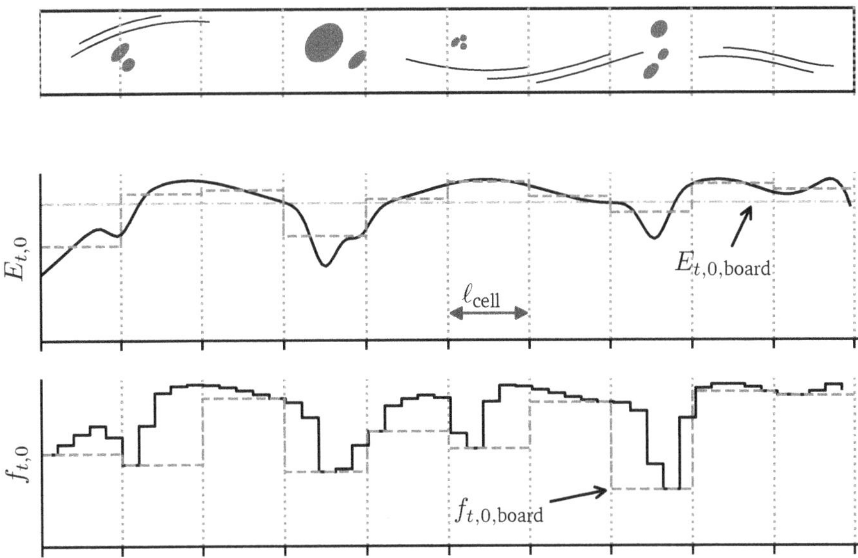

Figure 2.3. Illustration of the variation of mechanical properties along a lumber board and the effect of growth-bound defects (knots and grain deviation). The *continuum* variation of MOE and $f_{t,0}$ is represented by the solid lines in the upper and lower diagrams, respectively. The dashed line corresponds to the mechanical properties in discrete windows of length ℓ_{cell}.

shown by the solid line of the lower diagram of Fig. 2.3.

In reality, however, highly localized tensile strength profiles are impossible to determine—owed to obvious practical limitations—, and continuous tensile MOE profiles of boards were not practical to measure until rather recently, with the wider availability of different optical systems. Thus, the common approach to the study of "localized" mechanical properties of boards has been the consideration of discrete segments over which the different properties are measured (see dashed lines in Fig. 2.3), and build a mathematical model based on this information. The segments vary in size and regularity, depending on the model.

The literature presents numerous models that apply this concept for different properties (e.g. MOE and tensile or bending strength), varying in complexity and needed input variables. These models have, in many cases, been applied to existing glulam strength models (Foschi and Barrett, 1980; Ehlbeck and Colling, 1987; Showalter et al., 1987; Blaß et al., 2005), whilst others (to the knowledge of the author) were mainly used to investigate the size effect associated to the length of the boards (Lam and Varoglu, 1991b; Taylor and Bender, 1991).

In the following, different models developed for the segment-wise simulation of board properties are presented.

2.5.1 Predictor-based stochastic models

Regression models have been extensively used in material models for glulam simulations, i.a. by Ehlbeck et al. (1984), Colling (1990b), Blaß and Frese (2006) and Fink (2014). They are composed of two parts: one for the allocation of predictor parameters (like knot area ratio, KAR, and density ρ), and second part, where the mechanical properties are computed for each cell/segment, based on the previously generated values. The uncertainties are taken into account by means of a random component, associated to the standard deviation of the residual terms of the fitted equations. The evolution of this kind of model is presented here, starting with the original model from Foschi and Barrett (1980).

Foschi and Barrett (1980) developed a highly detailed material model for their glulam strength model (see Sec. 2.6.1). The approach is based on Weibull's weakest link theory and considers the effect of knots in a fairly detailed analytical manner. It contemplates the simulation of MOE (E_c) and tensile strength (f_c) values for clear wood segments by means of the following two-parameter Weibull model:

$$E_c = M_e f_g \left(\frac{G}{G_m}\right)^{1.25} [-\ln(1-p)]^{1/k_e} \tag{2.17}$$

$$f_c = M_s f_g \left(\frac{G}{G_m}\right)^{1.5} \left[\frac{-\ln(1-p_f)}{V}\right]^{1/k_s}, \tag{2.18}$$

where p is the probability level, M_s, M_e, k_s, k_e and f_g correspond to model parameters to be calibrated, and G and V are the density and volume, respectively. The MOE equation from Eq. (2.17) is modified according to the knot size diameter as

$$E_{t,0} = \alpha E_c, \tag{2.19}$$

where α is a fracture-mechanics-based factor that takes into account the effect of the knot, which for the sake of brevity is not presented in detail here (for the exact equation of α refer to Foschi and Barrett (1980)). The tensile strength, $f_{t,0}$, of each cell is subjected to the same correction for the case of pure tension. However, when the cell is considered as being part of a glulam beam, further modifications factors are applied to consider non-uniform loading (θ) and lamination effect due to the lateral restraining of deflections (ψ):

$$f_{t,0} = \alpha\theta\psi f_c. \tag{2.20}$$

The model then considers a randomly generated density G (obtained from a normal distribution $\mathcal{N}(G_m, \sigma_G)$) and a knot size D (obtained from frequency data) for each cell. Then, the values E_c and f_c are computed according to Eqs. (2.17) and (2.18) for a random probability (failure) level p (p_f), and the values MOE and tensile strength are then computed with Eqs. (2.19) and (2.20), respectively.

Ehlbeck and Colling developed a model for the simulation of MOE parallel to the grain ($E_{t,0}$) and tensile strength ($f_{t,0}$) by means of linear regressions, considering the knot area ratio (KAR) and density (ρ). According to their model, the properties are computed as

$$\ln(E_{t,0}) = \beta_0 + \beta_1 \mathrm{KAR} + \beta_2 \rho_0 + S(0, \sigma) \tag{2.21}$$
$$\ln(f_{t,0}) = \beta_0 + \beta_1 \ln(E_{t,0}) + \beta_2 \mathrm{KAR} \cdot \ln(E_{t,0}) + S(0, \sigma), \tag{2.22}$$

where the values β_i correspond to the parameters to be calibrated, KAR and ρ_0 are the knot area ratio and density at 0 % moisture content (MC), and the values $S(0, \sigma)$ correspond to the normally distributed residuals of the fitted regression. The calibrated parameters for these and the following models are given in Table 2.1. The model also contemplates regression equations for the compression MOE ($E_{c,0}$) and compressive strength ($f_{c,0}$), for which additionally the moisture content is regarded (see Ehlbeck et al. (1984), Table 3.2 for the rest of the regression equations and fitted parameters).

The KAR-values are taken from the respective statistical distribution and assigned to each 150 mm long cell along the boards. Density is assumed constant throughout each board with a value taken randomly form the corresponding statistical distribution. This means that the variation within clear wood (KAR = 0) is exclusively determined by the residual term $S(0, \sigma)$, as both predictors are constant for these segments (KAR = 0; ρ_0 = const.). This model was used by Colling (1990b) for his simulations of glulam of softwoods, too.

The adaptation of the "Karlsruher Rechenmodel" to beech glulam added modifications to the original regression equations (Blaß et al., 2005; Blaß and Frese, 2006), where the term ρ_0^2 (density squared) was included:

$$\ln(E_{t,0}) = \beta_0 + \beta_1 \mathrm{KAR} + \beta_2 \rho_0 + \beta_3 \rho_0^2 + S(0, \sigma) \tag{2.23}$$
$$\ln(f_{t,0}) = \beta_0 + \beta_1 E_{t,0} + \beta_2 E_{t,0} \cdot \mathrm{KAR} + S(0, \sigma). \tag{2.24}$$

These regressions were calibrated with a beech dataset (see Table 2.1), for which an R-value of 0.76 and 0.88 was obtained for $E_{t,0}$ and $f_{t,0}$, respectively (Blaß et al., 2005). Regression equations for $E_{c,0}$ and $f_{c,0}$ were developed, too, but are omitted here (for details refer to Blaß et al., 2005).

Experimental investigations by Fink (2014) led to the identification of the predictors with highest correlation with $E_{t0,}$ and $f_{t,0}$ for Norway spruce specimens. The variables found were the dynamic modulus of elasticity, $E_{\text{dyn,F}}$, (obtained from eigenfrequency measurements) and the total knot area ratio, tKAR (here refered to simply as KAR). Equations were developed for the MOE separately for clear wood segments (CWS) and weak sections (WS) as

$$\ln(E_{t,\text{CWS}}) = \beta_0 + \beta_1 E_{\text{dyn,F}} + S(0,\sigma) \tag{2.25}$$

$$\ln(E_{t,\text{WS}}) = \beta_0 + \beta_1 E_{\text{dyn,F}} + \beta_2 \text{KAR} + S(0,\sigma), \tag{2.26}$$

whilst for the tensile strength, $f_{t,0}$, equations for the entire board and for WS were fitted:

$$\ln(f_{t,\text{board}}) = \beta_0 + \beta_1 E_{\text{dyn,F}} + \beta_2 \text{KAR} + S(0,\sigma) \tag{2.27}$$

$$\ln(f_{t,\text{WS}}) = \beta_0 + \beta_1 E_{\text{dyn,F}} + \beta_2 \text{KAR} + S(0,\sigma). \tag{2.28}$$

Fink argues that the equation to simulate $E_{t,\text{WS}}$ (Eq. (2.26)) can be used in the CWS as well, using KAR = 0. However, for the case of $E_{t,0}$, this leads to an underestimation of about 3 % (Fink, 2014). The tensile strength of weak sections ($f_{t,\text{WS}}$) is computed with Eq. 2.28, which was calibrated by means of a linear regression analysis for censored data. This method was used to fill the missing values ($f_{t,\text{WS}}$) of the rest of the WS in each board (see details in Fink, 2014, Chapter 5). Similar as for the MOE, this equation—originally fitted using weak sections only—can be used for the simulation of CWS, too. However, the generated values $f_{t,\text{CWS}}$ are slightly underestimated in this case. The model implicitly considers a size effect due to the presence of knots: if the board is longer, then the probability of having a higher KAR value increases, which leads to a lower global $f_{t,0}$.

2.5.2 Pure stochastic models

An alternative approach to simulate the variability of mechanical properties within boards consists in the direct use of statistical information available for each variable.

Table 2.1. Summary of the presented regression models and their corresponding parameters for a given species, as reported in the respective literature

Species	Model	Parameters
Spruce	*Karlsruher Rechenmodel:* (Ehlbeck and Colling, 1987)	
	$\ln(E_{t,0}) = \beta_0 + \beta_1 \mathrm{KAR} + \beta_2 \rho_0 + S(0, \sigma)$	$\beta_0 = 8.2$ $\beta_1 = -1.17$ $\beta_2 = 3.13$ $\sigma = 0.180$
	$\ln(f_{t,0}) = \beta_0 + \beta_1 \ln(E_{t,0}) + \beta_2 \mathrm{KAR} \cdot \ln(E_{t,0}) + S(0, \sigma)$	$\beta_0 = -4.22$ $\beta_1 = 0.876$ $\beta_2 = -0.093$ $\sigma = 0.187$
Beech	*Karlsruher Rechenmodel:* (Blaß et al., 2005)	
	$\ln(E_{t,0}) = \beta_0 + \beta_1 \mathrm{KAR} + \beta_2 \rho_0 + \beta_3 \rho_0^2 + S(0, \sigma)$	$\beta_0 = 3.36 \times 10^{-1}$ $\beta_1 = 2.64 \times 10^{-2}$ $\beta_2 = -1.56$ $\beta_3 = 1.87 \times 10^{-5}$ $\sigma = 0.182$
	$\ln(f_{t,0}) = \beta_0 + \beta_1 E_{t,0} + \beta_2 E_{t,0} \cdot \mathrm{KAR} + S(0, \sigma)$	$\beta_0 = 3.09$ $\beta_1 = 9.76 \times 10^{-5}$ $\beta_2 = -1.54 \times 10^{-4}$ $\sigma = 0.239$
Spruce	*Model by Fink (2014):*	
	$\ln(E_{t,\mathrm{CWS}}) = \beta_0 + \beta_1 E_{\mathrm{dyn,F}} + S(0, \sigma)$	$\beta_0 = 8.52$ $\beta_1 = 7.12 \times 10^{-5}$ $\sigma = 5.47 \times 10^{-2}$
	$\ln(E_{t,\mathrm{WS}}) = \beta_0 + \beta_1 E_{\mathrm{dyn,F}} + \beta_2 \mathrm{KAR} + S(0, \sigma)$	$\beta_0 = 8.41$ $\beta_1 = 7.69 \times 10^{-5}$ $\beta_2 = -9.02 \times 10^{-1}$ $\sigma = 1.0 \times 10^{-1}$
	$\ln(f_{t,\mathrm{board}}) = \beta_0 + \beta_1 E_{\mathrm{dyn,F}} + \beta_2 \mathrm{KAR} + S(0, \sigma)$	$\beta_0 = 2.14$ $\beta_1 = 1.13 \times 10^{-4}$ $\beta_2 = -1.08$ $\sigma = 2.77 \times 10^{-1}$
	$\ln(f_{t,\mathrm{WS}}) = \beta_0 + \beta_1 E_{\mathrm{dyn,F}} + \beta_2 \mathrm{KAR} + S(0, \sigma)$	$\beta_0 = 2.96$ $\beta_1 = 8.5 \times 10^{-5}$ $\beta_2 = -2.22$ $\sigma = 1.5 \times 10^{-1}$

This type of models are not intended to estimate the specific properties of a board—as it is the case of the previously presented models—, but rather reproduce the observed general behavior of a given population of boards. For this, a commonly used concept is the serial correlation, or autocorrelation, which indicate the degree of variation from one position on the board to the next.

The first application of the serial correlation concept to simulate the variation of MOE along boards was developed by Kline et al. (1986), where a second order AR process (see Section 2.4.3) was used to simulate the properties of South pine lumber in segments of 762 mm (30 in) in length. The model proposed by Kline et al. (1986) takes the following form:

$$E_i = \varphi_1 E_{i-1} + \varphi_2 E_{i-2} + \varepsilon_i, \tag{2.29}$$

where the random term ε_i is defined as follows (if data normality is assumed)

$$\varepsilon_i = \sigma_E \cdot t_i \cdot \sqrt{1 - R^2}. \tag{2.30}$$

Here, t_i corresponds to a random observation from a normal distribution $\mathcal{N}(0, 1)$, σ_E is the standard deviation of the observed MOE data, and R is the coefficient of correlation resulting from fitting the experimental data to Eq. (2.29). This model was calibrated with data obtained from bending tests in adjacent segments of length equal to 762 mm.

Showalter et al. (1987) applied the same principles to the development of a strength model, after noticing that serial correlation models gave better results for the simulation of tensile strength. The proposed model considers two parallel Markov processes: (i) one being the MOE simulated according to Kline et al. (1986) (Eq. 2.29), (ii) and the other coming from the serial correlation of the residuals of the tensile strength. The residuals, are computed according to the first order Markov process as

$$\varepsilon_i = r_1 \frac{\sqrt{k \cdot E_i}}{\sqrt{k \cdot E_{i-1}}} \cdot \varepsilon_{i-1} + t_i \sqrt{k \cdot E_i} \cdot \sqrt{1 - r_1^2}, \tag{2.31}$$

where the product $(k \cdot E_i)$ represents the variance of the tensile strength at the segment i, k is a model parameter, and r_1 is the lag-1 correlation of the strength residuals. This model assumes then that the standard deviations of the residuals of the MOEs (represented by the term $\sqrt{k \cdot E_i}$) vary for each segment, making it a non-stationary process.

Both AR models are then used in the following weighted least squares regression model to obtain the strength at each segment:

$$f_{t,i} = \exp(\beta_0 + \beta_1 E_i + \varepsilon_i) \quad , \tag{2.32}$$

where β_i are model parameters, E_i is obtained with Eq. (2.29), and ε_i is computed according to Eq. (2.31).

Taylor and Bender (1988) extended the previous concept to allow for the simulation of correlated, non-normally distributed variables. It was originally used for the simulation of modulus of elasticity and strength values for lumber. The developed procedure uses the multivariate normal distribution to generate correlated data according to a covariance matrix $\mathbf{\Sigma}$. The random variates are then mapped to the statistical distributions corresponding to each random variable by means of the inverse of the cumulative distribution function. The method can be described as follows:

1. The random variables $X_1, X_2, \ldots, X_N$ are fitted to suitable distribution functions $F_{X_1}, F_{X_2}, \ldots, F_{X_N}$.
2. The covariance matrix, $\mathbf{\Sigma}$, for the different variables is computed according to

$$\mathbf{\Sigma} = \begin{bmatrix} \sigma_{X_1}^2 & \sigma_{X_1,X_2} & \cdots & & \sigma_{X_1,X_N} \\ & \sigma_{X_2}^2 & \sigma_{X_2,X_3} & \cdots & \sigma_{X_2,X_N} \\ & & \sigma_{X_3}^2 & \cdots & \sigma_{X_2,X_N} \\ & \text{Sym.} & & \ddots & \vdots \\ & & & & \sigma_{X_N}^2 \end{bmatrix} , \tag{2.33}$$

   where $\sigma_{X_i}^2$ is the variance of X_i (for $i = 1, 2, \ldots, N$) and σ_{X_i,X_j} is the covariance of the random variables X_i and X_j (for $i, j = 1, 2, \ldots, N$, $i \neq j$).
3. Generate the needed number of random variates Y_i using a multivariate normal distribution and the covariance matrix $\mathbf{\Sigma}$ (see Sec. 2.4.1).

4. Evaluate the cumulative distribution function $\Phi(y_i)$ for each generated value in each random vector. This generates correlated random vectors, $\mathbf{U}_i$, uniformly distributed between 0 and 1 ($U(0,1)$).

5. Apply the inverse of the cumulative distribution function $F_{X_i}^{-1}$ to the corresponding values u_i computed in the previous step. This yields the correlated vectors $\underline{\boldsymbol{x}}_i$ correlated according to $\boldsymbol{\Sigma}$.

The correlation between the variables is inherited from the observations generated from the multivariate normal distribution, and are thus expressed in the generated vectors $\underline{\boldsymbol{x}}_1, \underline{\boldsymbol{x}}_2, \ldots, \underline{\boldsymbol{x}}_N$, too.

Taylor and Bender (1991) applied their previously developed method to simulate correlated global properties of boards to the simulation of properties within Douglas-fir boards on 2 ft long segments. To achieve this, a vector of length $2N$ is generated, where N is the number of segments for which material properties should be simulated; The first N terms represent the MOE and terms $N+1$ to $2N$ are the tensile strength values:

$$\mathbf{X} = \begin{array}{c} \\ \end{array} \overset{\begin{matrix} 1 & 2 & \dots & N & N+1 & N+2 & \dots & 2N \end{matrix}}{\begin{bmatrix} E_{t,0,1} & E_{t,0,2} & \dots & E_{t,0,N} & f_{t,0,1} & f_{t,0,2} & \dots & f_{t,0,N} \end{bmatrix}} \tag{2.34}$$

The covariance matrix $\boldsymbol{\Sigma}$ is in this case a $2N \times 2N$ matrix, where the serial and cross-correlations of both properties are specified:

$$\boldsymbol{\Sigma} = \begin{matrix} \\ 1 \\ 2 \\ \vdots \\ N \\ N+1 \\ \vdots \\ 2N \end{matrix} \begin{matrix} \begin{matrix} 1 & 2 & \cdots & N & N+1 & \cdots & 2N \end{matrix} \\ \begin{bmatrix} \sigma_{E_1}^2 & \sigma_{E_1,E_2} & \cdots & \sigma_{E_1,E_N} & \sigma_{E_1,f_1} & \cdots & \sigma_{E_1,f_N} \\ & \sigma_{E_2}^2 & \cdots & \sigma_{E_2,E_N} & \sigma_{E_2,f_1} & \cdots & \sigma_{E_2,f_N} \\ & & \ddots & \vdots & \vdots & \ddots & \vdots \\ & & & \sigma_{E_N}^2 & \sigma_{E_N,f_1} & \cdots & \sigma_{E_N,f_N} \\ & & & & \sigma_{f_1}^2 & \cdots & \sigma_{f_1,f_N} \\ & & \text{Sym.} & & & \ddots & \vdots \\ & & & & & & \sigma_{f_N}^2 \end{bmatrix} \end{matrix} \tag{2.35}$$

Thus, the problem is transformed into a simulation of 2N correlated variables, where the first N variables and the last N variables follow the statistical distributions corresponding to the modulus of elasticity and tensile strength, respectively.

Although serial-correlation data for lags higher than three were not available (specimens were composed of four segments), these values were computed and included according to the autocorrelation function (ACF) of and AR(3) model for the case of the MOE, according to

$$\rho_k = \beta_1 \cdot \rho_{k-1} + \beta_2 \cdot \rho_{k-2} + \beta_3 \cdot \rho_{k-3}, \tag{2.36}$$

where ρ_k is the lag-k. An AR(1) was assumed for the case of the tensile strength (where only the lag-3 correlation was known)

$$\rho_k = \rho_3^{k/3}. \tag{2.37}$$

Lam and Varoglu (1991b) presented a serially correlated model for the tensile strength on 610 mm long segments of Spruce-Pine-Fir specimens. Based on experimental results, a Moving Average process of order 3 [MA(3)] was developed (Lam and Varoglu, 1991a). The method consists on two stages: in a first step, (1) uncorrelated data taken from a log-normal distribution is generated for each 610 mm segment ($\widetilde{f}_{t,i}$), then (2) the data is correlated with the MA model. The parameters for the log-normal distribution (mean and COV) are described as functions of the minimum tensile strength of each board ($f_{t,\min}$), which in turn is generated from a two-parameter Weibull distribution. The mean value, μ, of the uncorrelated data is then used in the MA model as:

$$f_{t,i} = \mu + \sum_{j=1}^{q=3} a_j \varepsilon_{i-j}, \tag{2.38}$$

where the parameters a_i are the weighting factors of the moving average model. The values ε_i correspond to the difference between the strength data generated in the first step, $\widetilde{f}_{t,i}$, and their mean μ, i.e. it is the variation around its mean value:

$$\varepsilon_i = \widetilde{f}_{t,i} - \mu. \tag{2.39}$$

Monte-Carlo simulations of the boards with different lengths performed by Lam and Varoglu (1991b) show a size effect comparable to that of experimental results.

A different approach was presented later by Lam et al. (1994), where a method to simulate continuous MOE profiles was introduced based on the analysis of bending profiles of spruce-pine-fir boards, obtained with a continuous bending grading machine (Cook-Bolinder type). The method was based on the previous work performed by Bechtel (1985), Foschi (1987) and Lam et al. (1993), where the localized bending MOE was approximated by means of a Fourier series. The rather long boards of 4.9 m studied in Lam et al. (1994) presented a clear trend in the MOE along the length of the board, making it a nonstationary process. Once the trend was removed, the process was assumed to be stationary and normally distributed.

The simulation of MOE profiles was composed of two parts: (i) the generation of a stationary process by means of a sum of cosines with empiric spectral amplitudes and (ii) the addition of a trend component to reproduce the observed nonstationarity. The key insight of this approach is the understanding that for the usual autoregressive analyses to work—including in this case the Fourier analysis—a stationary process is needed.

Isaksson (1999) developed a stochastic bending strength model considering weak and strong sections of Norway spruce. Here, the distance between consecutive weak sections is represented by a gamma distribution and the length of each weak section can be either constant (150 mm), randomly generated from a from a beta distribution, or equal to the sum of half the distance between two adjacent weak sections (no strong sections assumed in this case). However, the influence of these three options showed no noticeable effect in the results of simulations to study load and length effects (Isaksson, 1999). The bending strength of the strong sections is considered to be equal to the strongest weak section of the board. The strength values are generated for each weak segment j of a board i as

$$f_{t,i,j} = \exp(\mu + \delta_i + \varepsilon_{i,j}), \tag{2.40}$$

where μ is the mean strength of all weak sections in all boards, δ_i represents the difference between the mean μ and the mean of the weak sections in the board i, and $\varepsilon_{i,j}$ is the variation with respect to the mean value of the board. The values δ_i and $\varepsilon_{i,j}$ are obtained from normal distributions of zero mean and standard deviations σ_i and σ_j, respectively. The serial correlation was determined to be $R = 0.54$ and is equal for all the five investigated lags. To preserve it, the terms σ_i and σ_j must satisfy the following condition:

$$R = \frac{\sigma_i^2}{\sigma_i^2 + \sigma_j^2}. \tag{2.41}$$

The size effect is captured by generating 5.5 m long boards and then *cutting* them down to the needed size.

2.5.3 Simulation of KAR values along the boards

As some of the mentioned models for MOE and strength along boards depend on the local KAR information, simulation models for KAR along the board have been developed for this purpose, i.a. by Ehlbeck and Colling (1987), Isaksson (1999), and Blaß et al. (2005). Ehlbeck and Colling (1987) considered the KARs to be multiples of a per-board *characteristic* KAR ($\text{KAR}_i = \text{KAR}_{\text{char}} \cdot k_i$), for which a statistical distribution was determined. The variable k_i was allowed to move between the limits k_{min} and k_{max} (different for every board). The simulation starts by sampling a random variate from the KAR_{char} distribution and then generating a value k_i for each KAR that needs to be generated in the board, respecting the limits k_{min} and k_{max}. The distances between the knot-affected cells was determined by means of a similar process, where the distance between two adjacent knot-affected cells, ℓ_i was determined as $\ell_i = \ell_{\text{char}} \cdot t_i$. Here, t_i is bounded by the limits t_{min} and t_{max}.

The model implemented by Blaß et al. (2005) (explained in detail by Frese (2006b)) is based in the work done by Görlacher (1990), where a KAR_{max} value is assigned to each board, and is then multiplied by a factor between $0 < k_i \leq 1$ to obtain the rest of the KAR ratios. No special consideration regrading the distances between knot-affected cells was done, meaning that the positions of the knot-affected cells was randomly selected for each KAR value.

$$\text{KAR}_i = \text{KAR}_{\text{max}} \cdot k_i. \tag{2.42}$$

In his study, Fink (2014) described the KAR values as a hierarchical model with two levels: one for the between board variation, and the second one for the within board variation:

$$\text{KAR}_{i,j} = \exp(\mu + \tau_i + \varepsilon_{i,j}), \tag{2.43}$$

with

$$\tau_i + \varepsilon_{i,j} \leq \ln(\mathrm{KAR}_{\mathrm{limit}}) - \mu. \tag{2.44}$$

Compared to the model by Blaß et al. (2005), Fink considers a board length size effect intrinsically, as the probability to produce larger KAR values (larger $\varepsilon_{i,j}$ in Eq. (2.43)) increases with the number of generated KAR values. In contrast, taking the $\mathrm{KAR}_{\mathrm{max}}$ value per board requires the $\mathrm{KAR}_{\mathrm{max}}$ distribution to be *manually* adapted for different lengths. The model presented by Ehlbeck and Colling (1987) considers a length size effect too, although not explicitly mentioned.

2.6 Probabilistic strength models for glulam beams

Glued laminated timber, being composed of boards with an inherent variation of mechanical properties, exhibits a marked variation in mechanical properties such as stiffness and, especially, strength. Experimental tests constitute the most direct method to describe the statistical characteristics of said properties. However, due to the pronounced costs associated to such tests, diverse computational models have been developed to simulate the mechanical behavior of GLT beams. Most of these models are based on Monte-Carlo simulations, consisting on a series of simulations, where different variables are varied for each configuration according to given statistical distributions. The method has been the standard in the numerical analysis of GLT beam elements since the 1980', where different models have made use of its concept. In the following, a review of the most relevant glulam strength models is presented.

2.6.1 Model by Foschi and Barrett (1980)

The first reported glulam strength model to consider the stochastic distribution of mechanical properties of laminations throughout the beam was introduced by Foschi and Barrett (1980). It consisted of a finite element (FE) model where laminations were divided into equally spaced *cells* of 6 in (152 mm), as shown in Fig. 2.4. The FE mesh considered each one of these segments as one element. Densities and knot diameters were assigned to each cell in a random manner, taking their values from experimental statistical distributions. Moduli of elasticity (MOEs) and strength data were generated according to Eqs. (2.23) and (2.24), and analogous equations for the compressive properties. Finger-joints were not considered, due to a lack of representative data. The failure criterion was based

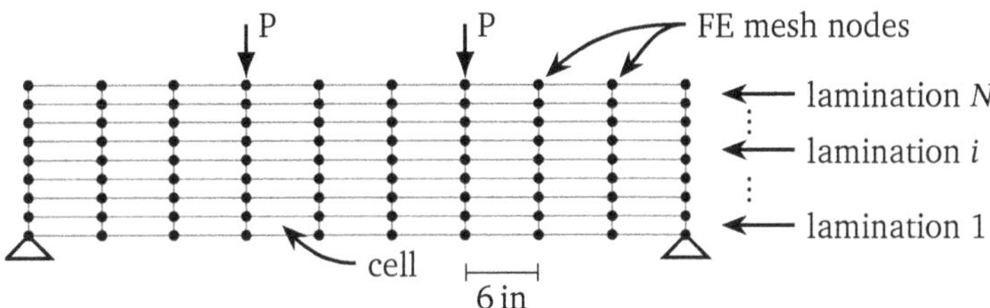

Figure 2.4. Mesh used by Foschi and Barrett (1980) for their finite element glulam strength model. Different mechanical properties are assigned to each element (cell)

on the weakest link assumption, i.e. the cell that fails first at a given applied load determines the ultimate load of the beam.

Although the results obtained from the model were satisfactory, the non-consideration of finger-joints left space for improvements. The reason for the omission of finger-joints in the analysis was mainly due to a lack of available data for finger-joints. Furthermore, no consideration regarding the spacial distribution of knots was made. Irrespective of this, the novel approach given by the stochastic nature of this model, and its potential for further enhancement, influenced the development of the models to come.

2.6.2 Karlsruher Rechenmodel

The "Karlsruher Rechenmodel" (*Karlsruhe computation model*) was initially developed by Ehlbeck et al. (1984), following the concept of defining localized material properties on equally sized *cells* of 150 mm along laminations, as done by Foschi and Barrett (1980). However, different as in previous models, the influence of the finger-joints was considered explicitly by assigning finger-joint material properties at cells joining two different boards. For this, regression equations relating the density of the jointed boards with their mechanical properties (MOE and tensile strength) were applied. The data for this came from an extensive experimental campaign. The distance between finger-joints is determined by the length of the boards, which are randomly generated from a statistical distribution determined by Larsen (1980). The mechanical properties of the timber elements were generated with regression equations based on the density and knot area ratio according to Eqs. (2.23) and (2.24), and analogous equations for the compressive properties.

A dedicated finite element program was implemented, where the mesh was built based on the principle of one element per cell, as done by Foschi and Barrett (1980). Plasticity was taken into account for the laminations subjected to compressive stresses (bending compression zone), and a refined failure criterion allowed for the failure of more than one cell before the ultimate load was achieved. The criterion was defined in the following manner: after one cell reaches its assigned strength at a given applied load ($P_{u,1}$) the element is removed ($E_{t,0} \approx 0$) and a new computation of the model is carried out. If the next cell to fail does so at a lower load than the first one ($P_{u,2} < P_{u,1}$), then the first load is taken as the ultimate load of the beam (crack propagation is assumed); if, on the other hand, the next cell fails at a higher load ($P_{u,2} \geq P_{u,1}$), then this cell is also removed and the previous condition is checked again. Thus, this criterion is able to account for the redistribution of internal load sharing between the different laminations after one of them locally fails (so-called *lamination effect*) and represents an important addition in comparison with previous models.

Improvements to the model were made a few years later by Ehlbeck and Colling (1987), where new correlation equations for the generation of the material properties were developed (see also Colling, 1990b). Most important, a better representation of the size and distribution of knots (in terms of the knot area ratio, or *KAR*) was implemented. The failure criterion was extended to include, in addition to the previous one, the following three points: (i) failure of a finger-joint located in the lowest lamination (tension zone) would immediately deliver the ultimate load of the glulam beam (no redistribution of stresses allowed); however, (ii) the failure of a finger-joint in any other lamination would be allowed, as long the stress redistribution disregarding the failed element leads to a further increase in the ultimate load. Finally, (iii) the failure of a board element at a knot location does not lead immediately to failure, if the beam showed enough reserve (capacity of redistributing the stresses).

This model is the current benchmark when talking about glulam strength models. The results obtained with it have helped in the development of the current state of the European standardization for the determination of characteristic strength values of glued laminated timber elements in EN 14080 (2013), and updates at more-or-less regular intervals ensure its validity in time.

2.6.3 PROLAM model

In the early 1990's a new model was proposed by Hernandez et al. (1992), which instead of using the finite element method, applied a "transformed section" approach. This method divides the glulam beam in segments of 2 ft (610 mm) along its length and transforms the width of each lamination in the composite cross-section, so that basic beam theory can be used on it. Similar as in the "Karlsruher Rechenmodel", the position of finger-joints is determined by randomly generated board lengths, and their mechanical properties are generated according to regression equations.

The material properties (MOE and $f_{t,0}$) varied within each board composing the glulam according to the model proposed by Taylor and Bender (1991), where the spacial correlation and cross-correlation of both variables is considered.

The failure criterion was similar to that presented originally by Ehlbeck and Colling (1987), where more than one cell was allowed to fail, until no further increase in the ultimate load (moment) was possible. This was done by analyzing each cross-section segment along the board and looking for the one where one of the laminations would first reach its assigned tensile strength—the stresses were measured at the mid-height of each lamination.

The proposed computational method (transformed sections) can deliver results in a much shorter time, as compared to the finite element method (especially for large cross-sections), but the gain in speed comes at the cost of considering less (or no) interaction between the different cells. In other words, since cross-sections segments are analyzed independently, there is no manner in which the effect of the adjacent segments can be taken into account for the computation of stresses in each cell. The results obtained with the model tend to underestimate the characteristic bending strength by 5 % and overestimate the MOE of the beam by 14.5 %, for which correction factors are applied. Nevertheless, no further studies are presented to show whether the correction factors apply to different configurations as well.

2.6.4 Karlsruher Rechenmodel (latest implementation)

In the context of the mentioned change in the growing stock of European forests in favor of hardwoods species, an adaptation of the "Karlsruher Rechenmodel" to glulam beams made of beech wood (*Fagus sylvatica*) was developed (Frese, 2006b; Blaß et al., 2005). To this end, an extensive experimental campaign was designed in order to obtain the needed statistical data for the material properties of boards and finger-joints, including statistical distributions and diverse correlations. In general, most of the original aspects of the model were preserved, e.g. random generation of board lengths (finger-joint separation), material properties variation along boards or the meshing used (see Fig. 2.4). However, some minor adjustments were made to the form of the regression equations for the material properties.

Further changes came on the software side, where the originally in-house developed FE-software was dropped in favor of the commercial software Ansys. The reimplementation of the model also came with a redefinition of the failure criterion, now defined as the load at which any cell located at the tensile outer edge of the beam fails—it is not clear whether failure of cells in e.g. the second lamination was allowed. Thus, this criterion does not consider the possible redistribution of internal forces after local cell failures occur as previous versions did. For this reason according to Blaß et al. (2005) the obtained results can be regarded to be conservative.

2.6.5 Model by Fink (2014)

Fink (2014) presented a strength model that differs in many aspects to the previously presented, while still maintaining the general principles. One of the differences consists in the used material model, which is centered in the differences between clear wood and weak sections (see Eqs. (2.25) to (2.28)). The material properties are then generated accordingly. The variation along the boards is mainly determined by the presence of knots (KAR-values), and the properties of segments between weak sections remain constant. Finger-joints are defined as segments with KAR-values between 0.2 and 0.3.

A FE-program was written in Matlab and the same mesh used in the previous models was applied (one element per cell/segment, see Fig. 2.4). The material was defined as isotropic, arguing that for the predominant bending situation no large errors are introduced. No ductility on the bending compression zone was considered. The failure criterion contemplated the possibility of more than one cell failing before ultimate load was reached. This was done by extending the original failure criterion of the "Karlsruher Rechenmodel" to account for noticeable changes in bending stiffness. After a cell has failed and the loads $P_{u,1}$ and $P_{u,2}$ are compared (see Sec. 2.6.2) a further comparison is made for the case when $P_{u,2} \leq P_{u,1}$. Namely, the bending stiffness of both systems (with and without removing the first failed element) are compared. If the new stiffness decreases by less than 1 %, then $P_{u,2}$ is accepted, otherwise $P_{u,1}$ is taken.

The model was validated with a set of experiments on glulam beams, where the location of each knot cluster and finger-joint, as well as E_{dyn}, were known. A very good agreement with the experiments was observed.

This model was extended by Blank et al. (2017) to include fracture mechanics by means of the consideration of a fracture energy, G_f, for the wood material. (Note: the model does not contemplate finger-joints explicitly, but rather they are considered as a wood section with a constant KAR value, which is calibrated to deliver strength values meeting the requirements specified in EN 1194 (1999).) For this, a *smeared crack* concept was applied, where the damage was descried by a stress-displacement relation in order to minimize mesh dependencies. Satisfactory results were obtained with a fracture energy $G_f = 10\,\text{N/mm}$, which was back-fitted to the experimental results. The rather high G_f value was justified by arguing that this fracture energy, representing the behavior of a smeared crack, incorporates additional effects as those observed in normal experiments of small, clear wood specimens.

2.6.6 Model by Kandler et al. (2018)

More recently, Kandler and Füssl (2017) presented a model to study the effects of the variability of wood on the linear behavior of GLT beams. Although, strictly speaking, this model does not qualify as a strength model—no prediction of failure is made—, two main aspects make it a very interesting model: Firstly, the model considers the generation of MOE data for boards based on the Karhunen-Loève expansion (Stefanou and Papadrakakis, 2007), which was calibrated on the base of high-resolution laser scans of fiber orientation in a grid of 1.2×4 mm (Kandler et al., 2015b). Secondly, and more relevant, it implements three different stochastic finite element methods: (1) the usual Monte-Carlo simulation, (2) the perturbation approach, and (3) the polynomial chaos projection scheme. The last two methods entail the potential of greatly reducing the computation time needed to obtain the same reliability as with the Monte-Carlo approach.

However, the consideration of advanced failure mechanisms under these simulation schemes is not trivial to implement, thus severely limits the practical applicability of the model. It also does not consider finger-joints, which in many cases is an crucial factor in the mechanical behavior of GLT beams, e.g. in hardwoods of oak, as shown by Aicher and Stapf (2014).

CHAPTER 3 MATERIALS

3.1 General remarks

As mentioned above, this thesis pursues two main objectives, (i) the study of the variation of properties within oak boards, and (ii) the development of a finite element based glulam strength model. Here, the results of (i) used in (ii) in the form of a mathematical model that accurately represents the material variation. In order to calibrate and validate both models, experimental data are needed, which, ideally, should stem from the same material sample. In this case, however, three different material datasets had to be employed employed: (1) Dataset A is used for the investigations on material properties variation throughout the boards, whilst (2) Dataset B is used for the validation of the glulam strength model; (3) the third dataset (C) is used to assess the applicability of the previously obtained parameters to a different dataset. All samples correspond to oak (*Q. robur*, *Q. petraea*) originating from France.

Since all datasets correspond to the same tree species and origin region, their growth-bound characteristics, such as knots and fiber orientation (as well as the properties influenced by these, e.g. MOE and $f_{t,0}$) should be comparable. Therefore, the simulation model for the variation of material properties along the boards, developed with the dataset A, is assumed to be suitable to simulate the board properties in the glulam strength model, which is validated with the dataset B. The following sections present the mechanical characterization of these datasets.

Table 3.1. Grading results for KAR and slope of grain for all investigated boards, and allocation to structural grades specified in DIN 4074-5 (2008)

	Grading limits according to DIN 4074-5			Grading results (mean ± std.)			
Grade	KAR-S (single)	KAR-G (group)	slope of grain	% of boards*	KAR-S (single)	KAR-G (group)	slope of grain
LS7	1/2	2/3	16%	8.5	0.21 ± 0.16	0.28 ± 0.18	14.12 ± 0.63
LS10	1/3	1/2	12%	36.2	0.14 ± 0.09	0.19 ± 0.11	7.30 ± 2.24
LS13	1/5	1/3	7%	46.8	0.10 ± 0.05	0.15 ± 0.09	3.52 ± 1.91

* The missing percentage (8.5 %) corresponds to the rejected boards

3.2 Dataset A: Material used for the study of within board variation

The investigations on the variation of grading parameters, e.g. density, knots, modulus of elasticity and tensile strength, were performed in a batch of oak boards, originating from the south-western part of France. The sample consisted of 52 boards and contained a mixture of appearance grades QF2 and QF3, according to EN 975-1 (2009). The nominal dimensions (length ℓ × width b × thickness t) of the planed boards were 2500 × 175 × 24 mm, respectively; the moisture content measured with a pin-type resistance meter was in average 10.2 % (COV = 4.6 %). The different QF-appearance grades intentionally allowed for a wide range of knotiness and grain deviation in the sample, leading presumably to a high variation in the properties of the respective boards. The boards were then visually graded at the MPA, University of Stuttgart, in hardwood strength grade classes LS7, LS10 and LS13 specified in the German structural hardwood grading standard DIN 4074-5 (2008), which conforms to EN 14081-1 (2016).

The strength grading with respect to knot-related variables and grain deviation was performed according to DIN 4074-5 (2008) with the needed variables obtained as described below. Great care was taken in order to allow for an exact quantification and allocation of the growth defects of each board. The results of the grading and defect quantification (KAR-values and slope of grain) are summarized in Table 3.1. The portions of the obtained different hardwood strength grades (LS7, LS10 and LS13) were: 8.5 %, 36.2 % and 46.8 %, respectively (8.5 % were rejected).

3.2.1 Segmentation of boards

Previous to the investigations on the variation of properties along boards, the central region of the specimens was divided into 15 equally-sized cells of $\ell_{cell} = 100\,mm$, as shown in Fig. 3.1. These segments (or cells) are used later to investigate the variation of material properties along the length of the boards. These segments are also used to study the relevant knot characteristics, such as inter-knot distances and frequency of knot sizes, which helps in the characterization of the studied material.

Different segment lengths have been used in previous similar studies; Ehlbeck and Colling (1987) and Blaß et al. (2005) used 150 mm, Kline et al. (1986) and Showalter et al. (1987) worked with 762 mm (30 in), and Lam and Varoglu (1991b) took 610 mm (2 ft) long segments, while Isaksson (1999) and Fink (2014) chose a variable length that depends on the inter-knot distances (clear wood). Such a diversity in studied segment lengths is not only a consequence of the preferred units systems of each author, but it depends on a variety of factors, such as the test configuration (bending or tensile tests), measurement technique, and the desired resolution. Furthermore, the species-dependent inter-knot distances are important. For this case, it was deemed desirable to obtain as many discrete values of the material properties throughout the boards as possible, given the measuring device available (see Chapter 5).

For the study of knot characteristics along the boards, four additional segments (also of length 100 mm) were considered at each side, i.e. before the cell No. 1 and after the cell No. 15, denoted by the superscripts "+/−" in Fig. 3.1. These

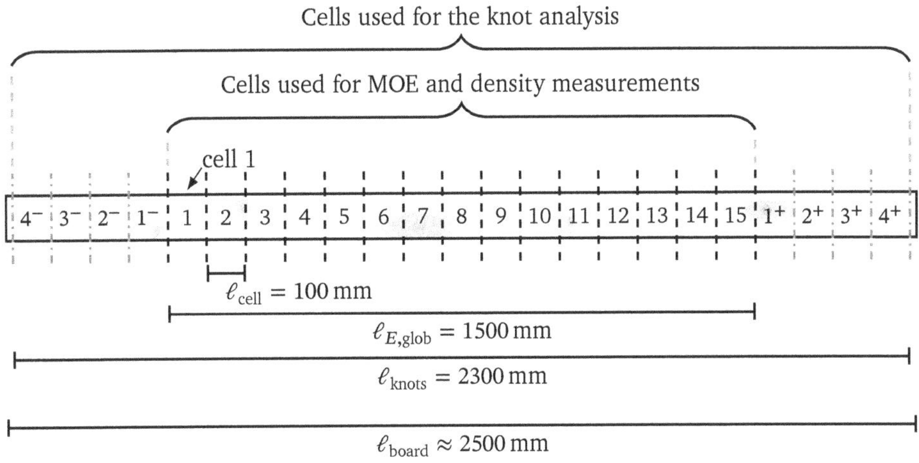

Figure 3.1. Illustration of the segmentation of the boards for the measurement of knot related variables and later measurement of MOE and tensile strength

additional cells help capturing the statistical characteristics of knots in the boards in a better, more reliable manner. A throughout analysis on the knot characteristics is presented in Chapter 4. The analysis of the mechanical properties corresponding to the central 15 cells is performed in Chapter 5.

3.2.2 Grain deviation

The maximum grain deviation along board length, i.e. the global slope of grain $\Delta y/\ell_g \cdot 100$, was measured manually for each board by means of a special scriber. The slope measurement was performed according to DIN 4074-5 (2008) with a base length of $\ell_g \approx 300\,\text{mm}$. The determination of the grain deviation stepwise along the board length is out of the scope of this investigation, but would certainly give additional, relevant information for the modulus of elasticity-tensile strength model predictions (see e.g. Kandler et al., 2015b or Olsson et al., 2018).

3.2.3 Knot-related variables

The position and dimensions of each knot larger than 5 mm was recorded using a digital caliper gauge. For each knot, three dimensions were measured: the minimum and the maximum diameter of the assumed ellipse, and the width taken perpendicular to the length axis of the board. These three variables allow for the later determination of the rotation angle of the knot, represented by an ellipse. These measurements were used to compute the KAR values according to DIN 4074-5 (2008). A detailed analysis of the knot variables along board is presented in Chapter 4.

3.2.4 Dynamic modulus of elasticity measurements

Eigen-frequencies measurements in the longitudinal direction were performed on each board previous to the testings with an impulse excitation measurement instrument of the type MK5i, from the company GrindoSonic. This device has a frequency range of 40 Hz–100 kHz and a precision better than 0.005 %. For the measurements, the boards were placed horizontally on a rubber-like surface to allow for an undisturbed vibration of the specimens, which were excited by means of a hammer of appropriate size. The dynamic MOE, E_{dyn}, was computed according to

Table 3.2. Statistical information on the measured dynamic MOE, E_{dyn}, separately for the different grades

Grade	N [–]	Mean [GPa]	Dynamic MOE, E_{dyn} Std [GPa]	COV [%]	Min. [GPa]	Max. [GPa]
Reject	4	10.4	1.1	10	8.9	12.0
LS7	4	11.3	2.6	23	8.1	15.1
LS10	17	12.2	2.1	18	8.9	16.4
LS13	22	12.1	2.0	17	9.2	17.0
All	47	12.0	2.1	18	8.1	17.0

$$E_{dyn} = 4 \cdot \ell_{board}^2 \cdot \nu^2 \cdot \rho, \tag{3.1}$$

where ℓ_{board} is the length of the board, ν is the eigen-frequency and ρ is the density. The measurement results are shown in Table 3.2 separately for each grading class and the whole sample.

3.3 Dataset B: Glulam experimental data: Gamiz

The material corresponding to Dataset B comprises white oak boards (*Q. robur*, *Q. petraea*) originating from the mid-eastern part of France (region Bourgogne Franche Comté), tested in the frame of a European Technical Approval for glulam beams of oak (ETA-13/0642, 2013) at the MPA, University of Stuttgart (Aicher and Stapf, 2014). The experimental campaign included tensile tests of boards, as well as tensile and bending tests of finger-jointed boards of cross-sections 100 × 20 mm and 140 × 20 mm. For the present study, only the boards with cross-sectional widths of 100 mm are considered.

Glued laminated timber specimens were produced from the same boards and were tested in bending, compression, tension and shear. More importantly, three different cross-sectional depths of 120 mm, 200 mm and 300 mm were produced. This allows for a later quantification of the size effect associated to the glulam made of this material. For each glulam beam the length of each board (distance between adjacent finger-joints) was recorded. These data have been previously analyzed and used for a simplified statistical glulam strength model by Aicher and Stapf (2014).

Table 3.3. Summary of the results for the tensile strengths of boards and finger-joints, and MOE parallel to the fiber of boards for grades LS10 and LS13 (Aicher and Stapf, 2014); Dataset B

Variable	Grade	N	Mean	Std	Min.	Max.	COV	x_{05}
$E_{t,0}$ [GPa]	LS10	10	12.7	2.0	8.7	15.9	16 %	8.7
	LS13	10	13.4	3.1	9.1	19.5	23 %	8.1
	LS10+LS13	20	13.1	2.6	8.7	19.5	20 %	8.7
$f_{t,0}$ [MPa]	LS10	50	45.1	14.3	23.1	83.8	32 %	24.6
	LS13	50	55.2	16.8	25.3	105.6	30 %	29.6
	LS10+LS13	100	50.1	16.4	23.1	105.6	33 %	26.8
$f_{c,0}$ [MPa]	LS13	14	50.1	2.3	46.0	55.1	5 %	45.3
$f_{t,j}$ [MPa]	LS10	49	44.6	9.9	26.5	66.9	22 %	29.0
	LS13	49	42.4	9.9	29.4	79.9	23 %	27.9
	LS10+LS13	98	43.5	10.0	26.5	79.9	23 %	28.8

3.3.1 Mechanical properties of boards

Boards corresponding to the grading classes LS10 and LS13 (DIN 4074-5, 2008) were tested. The mean density was 675 kg/m^3 and 641 kg/m^3 for grades LS10 and LS13, respectively. A total of 100 boards (50 LS10 and 50 LS13) were tested in tension, resulting in characteristic tensile strengths, $f_{t,0,k}$, of 24.6 N/mm^2 and 29.6 N/mm^2 for grades LS10 and LS13, respectively. For both grades, MOE measurements were performed on a total of 14 specimens (7 LS10 and 7 LS13) according to EN 408 (2012). The mean values obtained for $E_{t,0,\text{mean}}$ were 12.7 GPa and 13.4 GPa for grades LS10 and LS13, respectively.

For the finger-jointed boards, a total of 98 boards (49 LS10 and 49 LS13) were tested in tension, yielding characteristic tensile strength values of 29.0 N/mm^2 and 27.9 N/mm^2 for grades LS10 and LS13, respectively. The finger-joint profile employed is termed 10/3.8 (EN 14080, 2013), where the first number denotes the finger length and the second number relates to the pitch. The compressive strength parallel to grain, $f_{c,0}$, was only tested for grade LS13. Here, a characteristic value of $f_{c,0,k} = 45.3$ N/mm^2 was obtained. A summary of all the relevant statistics of the boards is presented in Table 3.3. Here, it is worth noticing the rather low values for the tensile strength of finger-joints as compared to the boards, which has direct consequences in the mechanical performance of the produced GLT beams, as the finger-joints will constitute the weak regions.

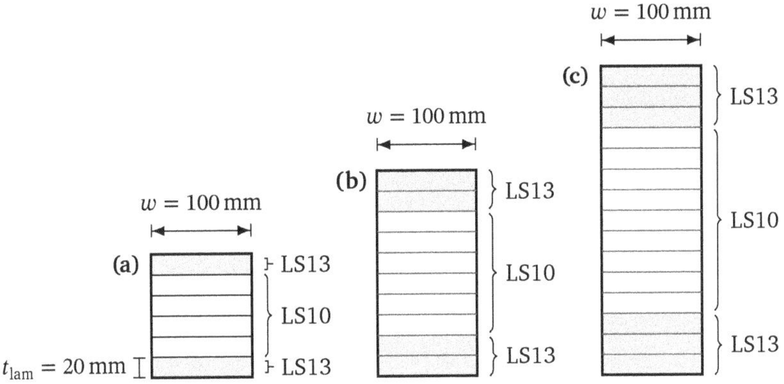

Figure 3.2. Cross-sections corresponding to the experiments of Table 3.4

Table 3.4. Mechanical properties measured for the glulam beams of dataset B. Characteristic values computed according to EN 14358 (2016)

Variable	Cross-section [mm²]	N	Mean	Std	Min.	Max.	COV [%]	x_{05}
$E_{m,g}$ [GPa]	100 × 120	10	10.8	0.8	9.2	12.3	8	9.1
	100 × 200	10	14.9	0.7	13.9	16.3	5	13.4
	100 × 300	10	14.7	0.8	13.2	15.7	5	13.1
$f_{m,g}$ [MPa]	100 × 120	10	54.4	5.4	48.5	63.3	10	43.7
	100 × 200	10	47.9	4.9	39.3	53.9	10	37.7
	100 × 300	10	43.4	4.3	35.0	48.6	10	34.3

3.3.2 Mechanical properties of glulam beams

Inhomogeneous glulam beams with LS10 boards in the inner part and LS13 boards in the outer regions of the beam were tested. The dimensions of the cross-sections (and the number of boards in each section) are: 100 × 120 mm (1-4-1); 100 × 200 mm (2-6-2); 100 × 300 mm (3-9-3). The details can be seen in Fig. 3.2.

The mean bending stiffness were 10.8 GPa, 14.9 GPa and 14.7 GPa for the depths 120 mm, 200 mm and 300 mm, respectively. Meanwhile, the characteristic bending strength, computed according to EN 14358 (2016), were 43.7 N/mm², 37.7 N/mm² and 34.3 N/mm² for depths of 120 mm, 200 mm and 300 mm, respectively. All the relevant statistics of the studied glulam beams can be taken from Table 3.4.

3.3.3 Distribution of board lengths in glulam

The length of the boards constituting the glulam beams was measured by determining the distance between adjacent finger-joints in each lamination. The results are presented in Table 3.5, where the rather short length of the boards is made evident ($\ell_{b,\text{mean}} \approx 500\,\text{mm}$).

Table 3.5. Statistics for the length of the boards constituting the glulam beams

N	Mean [mm]	Std [mm]	Min. [mm]	Max. [mm]	COV [%]	x_{05} [mm]
618	497	166	170	1280	33	277

3.4 Dataset C: FCBA experimental campaign

3.4.1 Mechanical properties of boards and finger-joints

Within the frame of the project "European hardwoods for the building sector" (EU Hardwoods, 2017), an exhaustive experimental campaign was performed on boards and GLT of oak by the research partners at FCBA, Bordeaux, France (Faye et al., 2017). The sample considered here consists on boards processed from logs of large diameters—excluding butt logs— and a cross-section of 27 × 160 mm. The moisture content fluctuated between 10 % and 12 %. Visual grading was done according to NF B 52-001-1 (2011), yielding mostly D24 grades (70 %), followed by D30 (24 %) and D18 (4 %), while 2 % was rejected. For the present study only the D24 grade is relevant, as that corresponds to the material used for the production of the here regarded GLT beams.

The MOE was determined for each board by means of dynamic excitation, using a mechanical timber grader. A total of 40 boards and 41 finger-joints, corresponding to the grade D24, were tested in bending following the specifications of EN 408 (2012). Due to lack of material, no tensile tests were performed on this set. The results are presented in Table 3.6 along with the results for compressive strength and MOE. However, tensile tests were later performed on a second sample of D24 boards, which should have similar characteristics as the first set. These results are shown in Table 3.6, too.

Table 3.6. Statistics for the tensile strengths of boards and finger-joints, and MOE parallel to the fiber of boards for the regarded sample of grade D24; Dataset B

Variable	Grade	N	Mean	Std	Min.	Max.	COV	x_{05}
$E_{dyn,0}$ [GPa]	D24	1637	11.5	2.1	6.0	19.9	18 %	8.4
$E_{m,0}$ [GPa]	D24	40	11.0	2.4	7.1	17.3	22 %	7.2
$f_{m,0}$ [MPa]	D24	40	80.4	20.0	46.2	124.9	25 %	48.5
$f_{t,0}$ [MPa][1)]	D24	30	31.2	10.5	10.3	59.2	34 %	14.3
$f_{m,j}$ [MPa]	D24	41	52.6	13.5	21.2	79.0	26 %	30.2
$f_{c,0}$ [MPa]	D24	20	46.3	2.6	40.0	50.7	6 %	41.4

[1)] Tensile tests performed on different dataset with similar characteristics (see 3.4.1)

Table 3.7. Mechanical properties measured for the glulam beams of dataset C. Characteristic values computed according to EN 14358 (2016)

Variable	Cross-section	N	Mean	Std	Min.	Max.	COV	x_{05}
$E_{m,g}$ [GPa]	160 × 160	20	11.2	0.9	10.1	13.0	8 %	9.6
	160 × 300	20	10.9	0.4	10.0	11.9	4 %	9.9
$f_{m,g}$ [MPa]	160 × 160	20	47.6	8.0	33.6	65.2	17 %	33.9
	160 × 300	20	40.8	4.5	34.2	50.7	11 %	32.7

3.4.2 Mechanical properties of glulam beams

The material corresponding to the D24 grade was used to produce a total of 40 GLT beams with two different cross-sections (width × height): 160 × 160 mm and 160 × 300 mm (eight and 15 laminations 20 mm in thickness, respectively); each configuration comprises 20 beams. Bending tests were performed according to EN 408 (2012), where a beam length of $18 \cdot d$ (d = beam depth) was used. The statistics of the results of both depths are given in Table 3.7. The statistics of the length of boards composing the GLT beams are presented in Table 3.8.

Table 3.8. Statistics for the length of boards used for the production of the tested GLT beams

N [–]	Mean [mm]	Std [mm]	Min. [mm]	Max. [mm]	COV [%]	x_{05} [mm]
1638	1832	400	1106	2826	22	1248

Analysis of knots in oak boards

4.1 General Remarks

A knot is caused by the inclusion of a branch in the wood of a tree stem (Kollmann and Côté, 1968), producing a highly localized fiber deviation around it. The detrimental effect of knots in the mechanical properties of timber boards is a known fact. This effect has been studied in numerous studies, mostly on softwoods (i.a. Isaksson, 1999; Foley, 2003; Fink, 2014; Olsson et al., 2013), but also for hardwood species, such as beech (Blaß et al., 2005). In general, regression analyses against some measure of knot size are used to quantify the influence of knots for modulus of elasticity (MOE) and strength. For the purpose of classification, simple regression analyses with global board properties are typically sufficient, as the localized properties are normally not needed. However, for the analysis and simulation of mechanical properties along board, additional parameters are usually considered, such as inter-knot distance or relative knot size within board.

The methods for the measurement of knot dimensions can be roughly divided in three groups: (i) direct measurement of visible knot dimensions, which can be done either manually or automatically by means of image processing algorithms; (ii) inference of knot dimensions by analyzing the orientation of fibers on the surface of the board, most commonly accomplished by means of laser scans that leverage the so-called *tracheid effect*; and (iii) the use of tomographic scans, which uses the differences in density of clear wood and knot material to reconstruct the geometry of knots.

For the characterization of knots of the oak boards studied here, the knot dimensions were manually measured and then digitally processed in order to estimate the true three-dimensional geometry. These data is then used to statistically analyze different knot-related parameters, which can be used to compare the knot characteristics of oak boards to those of previously studied species.

4.2 Knot area ratios (KAR)

The knot area ratio (KAR) is an indicating property (IP) used normally for the grading of boards according to DIN 4074-5 (2008). It gives a measure of the disturbance produced by a knot or a cluster of knots within a window of length L along the main axis of the board. It is defined as the projection of the knots within the mentioned window onto the cross-section of the board, as illustrated in Fig. 4.1. According to DIN 4074-5 (2008), the length shall be taken as $L = 150\,\text{mm}$, and the KAR must be evaluated at the most unfavorable section of each board. Formally, the KAR is defined as (DIN 4074-5, 2008):

$$\text{KAR} = \frac{\sum_{i}^{n} a_i}{2 \cdot b} \,, \tag{4.1}$$

where a_i are the measured knot projected dimension on the surface of the board, as shown in Fig. 4.1.

A directly related variable is the clear wood area ratio (CWAR), defined as

$$\text{CWAR} = 1 - \text{KAR}, \tag{4.2}$$

which is normally used for illustrative purposes, as it visually correlates better with changes in mechanical properties along boards, such as MOE. A drop in CWAR (higher KAR, more or larger knots) is normally associated with a drop in stiffness (see e.g. Isaksson, 1999). Furthermore, the KAR has been used in multiple analyses assessing the stiffness and strength properties on sections along boards by Ehlbeck and Colling (1987), Isaksson (1999), Blaß et al. (2005), and Fink (2014) (see

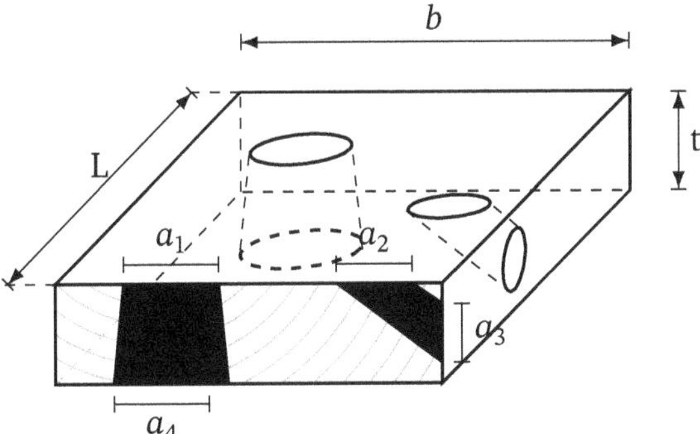

Figure 4.1. Determination of the knot area ratio (KAR) in a section of a board with length L. The black surface on the cross-section are the projections of the knots

Table 2.1).

4.2.1 Digitally reconstructing the 3D knot geometry of boards

Measurement of knots dimensions

The position and dimensions of each knot larger than 5 mm was recorded using a digital caliper gauge. For each knot, assumed to have an elliptical geometry, three dimensions were measured: the minimum and the maximum diameter of the assumed ellipse, and the width taken perpendicular to the length axis of the board. These three variables allow for the determination of the rotation angle of the knot. After this, the ellipses were plotted in a 2D graph at their respective position on the board, and the surfaces most probably corresponding to the same knot volume were manually assigned an identification number.

Reconstruction of knots

For the computation of the KAR values—both for grading purposes and assessment of property variation along board—a program was developed using the Python library pythonOCC (Paviot, 2018), which provides advanced 3D modelling functionality. The implemented algorithm can reconstruct the 3D geometry of each knot within each board, provided the position and dimensions of each knot on the four surfaces along the longitudinal axis of the board are given.

The algorithm used to reconstruct the knots is relatively simple compared to more advanced methods (e.g. Foley, 2003; Lukacevic et al., 2019) and can be summarized by the following steps:

1. The 3D geometry of the board is created as a rectangular box with sides $\ell_{\text{board}} \times b \times t$;
2. For each registered pair of surfaces belonging to the same knot, the corresponding ellipses are drawn at their respective location, and a volume passing through both ellipses is created (*lofting*);
3. For the case of knots defined only by a single surface on the board, a conic shape is assumed. Two cases are distinguished here: (i) if the knot occurs in a wide face, then the cone has a depth of $t/2$, (ii) if the knot is on a narrow face, then the cone has a depth $b/4$.

The third point represents a clear simplification of the real geometry of the knots, since the direction of the knot is in reality aligned with the LR-plane (longitudinal-radial) of the board directed towards the pith axis. Additional information regrading the pith location could be used to more accurately represent the direction of the cone simulating the knot, as done by Kandler et al. (2016). However, since the model will be used exclusively for the computation of KAR values, the errors that can be attributed to this simplification are expected to be small, as knots appearing only on one side are generally very small. The consideration of the pith location can be regarded for future improvements of the model, in order to better represent the knot orientation and allow for more complex analyses, such as the one presented in Lukacevic et al. (2019).

The *KAR* and corresponding clear wood area ratios (*CWAR*) were then numerically computed for each cell. This was done by projecting the knot volumes on the cross-section of each cell. The resulting areas were then used to compute the needed ratios.

Figure 4.2 illustrates a virtually reconstructed board using the described algorithm. The black surfaces on the front of each cell correspond to the clear wood area, whilst the white region corresponds to the area of the projected knot(s) within each cell. The latter area is obtained by firstly meshing the 3D surface of each knot (or part of knot) within each cell and then projecting the mesh elements onto the cross-section. For the cases where more than one knot lays within the same window, the projected areas are intersected to account for overlapping projections.

4.2.2 Comparison of different KAR definitions

As illustrated in Section 3.2.1, the studied boards were virtually subdivided in segments of 100 mm in length. The main purpose for this subdivision is the analysis

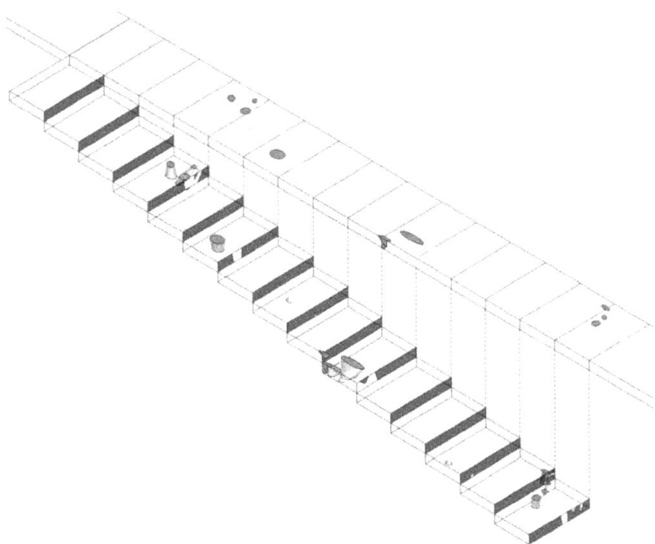

Figure 4.2. Post-processing of the gathered knot data, illustrated for board No. 50. Black areas in front of each of the cells correspond to the respective clear wood area ratio.

of localized MOE values along board (see Chapter 5). Since knots are known to negatively affect the MOE, it is reasonable to consider a knot variable in each cell, typically the KAR value, as this might help to explain the measured MOE variation. These data can then be used in conjunction with the MOE in further analyses.

However, the KAR values, measured in fixed discrete cells, do not satisfy the definition of KAR specified in DIN 4074-5 (2008) as (i) the used cell length is 100 mm instead of 150 mm and (ii) because the cells have a fixed location. The latter means that it is not assured that the regarded cells contain the largest possible KAR value in the board for the given window size. For example, if two adjacent cells have one knot each, then it might be possible to consider a shifted window that contains both knots, thus producing a higher KAR value. This argument was used by Fink (2014) to decide against the use of a fixed position of cells and consider a variable length for KAR and clear wood instead. Therefore, the KAR values obtained for the fixed cells of 100 mm in length have to be compared to the results obtained by a moving window that captures the maximum KAR value along the board. Also, the differences between using a window of 100 mm in length versus using a window of 150 mm in length should be assessed.

In the following, the KAR values computed at each cell, KAR_{cell} ($CWAR_{cell}$), are compared to the KARs obtained from a moving window of constant length along the board in a quasi continuous manner (steps of 10 mm). Here, two different window lengths are used: 100 mm, being the same size of the studied cells, and 150 mm, used i.a. by Ehlbeck and Colling (1987) and Blaß et al. (2005), and specified in DIN 4074-5 (2008) for grading purposes. The KAR (CWAR) values computed with these windows are denoted as KAR_{100} ($CWAR_{100}$) and KAR_{150} ($CWAR_{150}$), respectively.

Figures 4.3a–d show the three KAR definitions for a subset of five boards, displayed as CWAR for consistency with the analysis presented later in Chapter 5. It can be observed that the KAR_{cell} tend to underestimate the KAR of the weak zone, better represented by KAR_{100} and KAR_{150}. These two values show some differences as well, which is owed to the larger window length of the latter, allowing it to span more knots within its limits, thus yielding in general higher KAR values.

Comparisson of KAR_{cell} and KAR_{100}

The three KAR definitions are compared on the basis of *weak zones* (WZ), representing a region containing one or more knots. Specifically, a WZ starts when the KAR_{100} or KAR_{150} changes from zero to any value > 0, and ends when KAR_{100} or KAR_{150} returns to zero. The previous definition is illustrated in Fig. 4.4, where the identified WZs are represented by the shaded areas. The KARs to be compared correspond to the maximum values within the weak zones for each KAR definition (for instance, the red circles in Fig. 4.4 for KAR_{100}). The same concept is applied later to compare KAR_{100} and KAR_{150}.

Figure 4.5a shows the comparison between KAR_{cell} and KAR_{100}, where mostly a good agreement can be observed, denoted by approximately 85 % of the values matching almost exactly (difference under 5 %). This is illustrated in Fig. 4.5b, where the histogram of the relative difference between KAR_{cell} and KAR_{100} (termed $\Delta KAR_{cell\text{-}100}$) is presented, showing that 15 % of the KAR_{100} differ by up to around 40 % from the KAR_{cell} values.

For the case that only the maximum KARs of each board are considered, the percentage of exact matches slightly reduces to 70 %. This means that $\Delta KAR_{cell\text{-}100}$ tends to be larger when larger KARs are considered. At the same time, the maximum $\Delta KAR_{cell\text{-}100}$ decreases to 30 %.

Comparison of KAR_{100} and KAR_{150}

Figure 4.6a shows an analogous comparison for KAR_{100} and KAR_{150}, where a level of agreement of 78 % can be observed. Similar as before, Fig. 4.6b presents the histogram of $\Delta KAR_{100\text{-}150}$, showing a range reaching up to 60 %. If only the maximum KAR per board is considered, then the agreement is slightly reduced to 70 %, as well as the range of $\Delta KAR_{100\text{-}150}$ which drops down to 40 %.

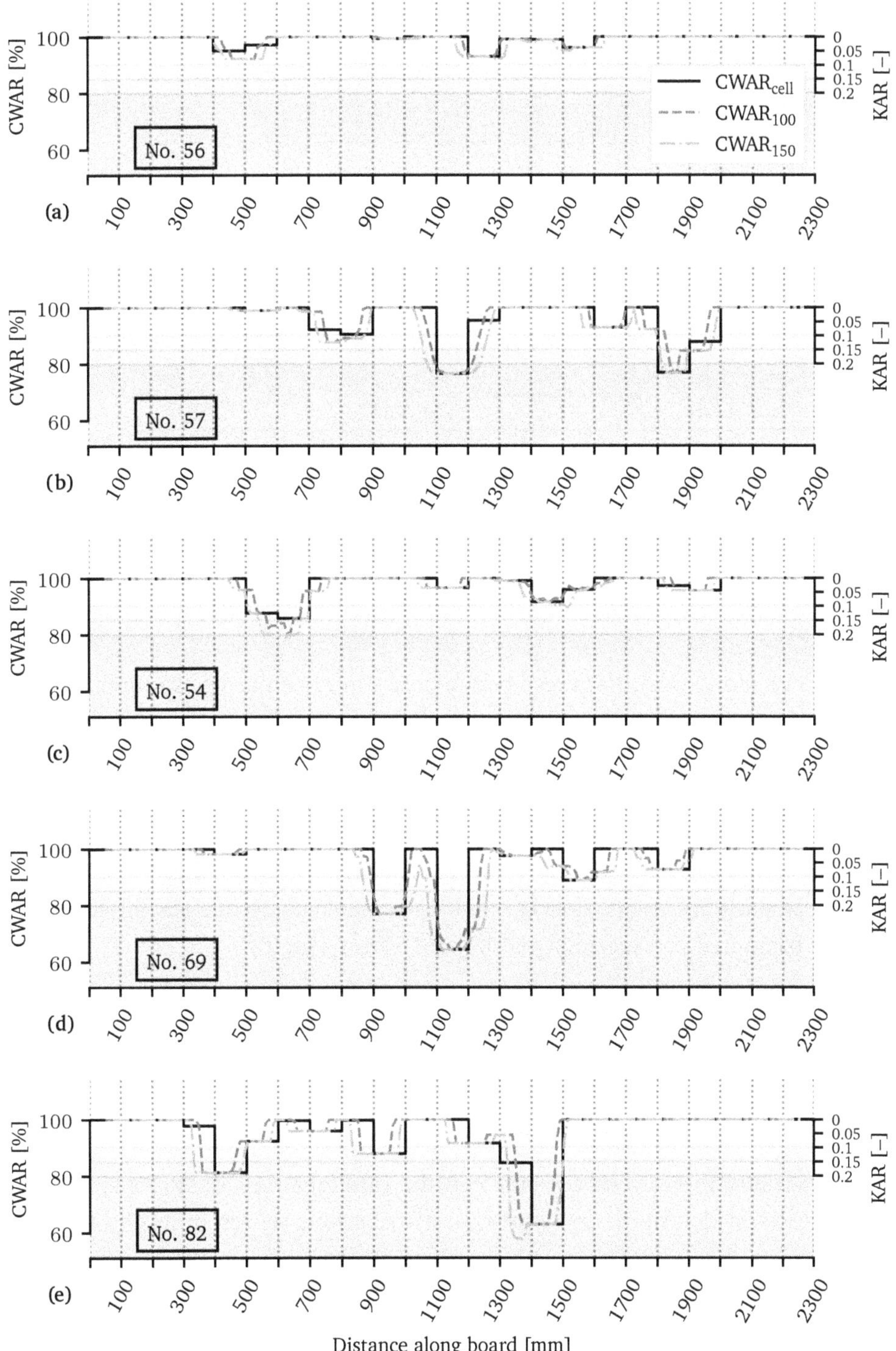

Figure 4.3. Comparison between different definitions of KAR values for a subset of the studied boards. Continuous line (100 mm fix): computation of KAR in windows of 100 mm at fixed steps of 100 mm; dashed line (100 mm cont.): computation of KAR in windows of 100 mm *continuously* over the board length; dash-dotted line (150 mm cont.): computation of KAR in windows of 150 mm *continuously* over the board length

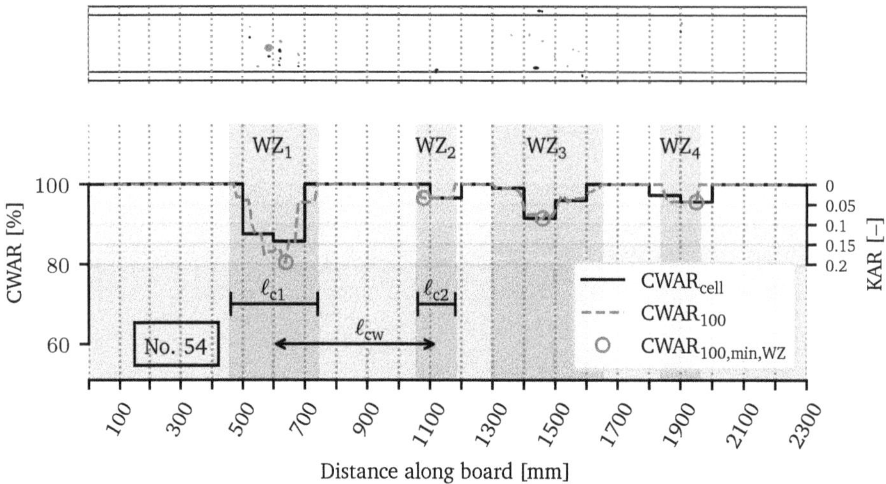

Figure 4.4. Definition of weak zones (shaded areas), $CWAR_{min}$ (KAR_{max}), and clear wood and weak zone distances

It has to be mentioned that these two analyses are comparing two different things: the former gives an insight into the effect of considering fixed windows of constant length for the KAR computation, and the latter compares the effect of the used window size. The analysis of the $\Delta KAR_{cell\text{-}100}$ in Fig. 4.5 indicates that the KAR_{100} should be preferred for simulation purposes due to the better representation of local maximum KAR values. The analysis of $\Delta KAR_{100\text{-}150}$ serves mostly for comparison reasons, as most of the KAR-related analyses made in similar studies are computed considering windows of 150 mm length.

4.2.3 KAR distributions

Absolute KAR values

Having illustrated the differences between the different KAR definitions, it is now of interest to determine the statistical distributions representing the KAR values. Although the KAR_{100} is to be the most appropriate variable for the purpose simulations, some the following analyses are performed for KAR_{cell} data as well. This helps in further illustrating the differences of considering KAR values obtained from adjacent windows of constant size versus a more flexible positioning of cells that capture the largest possible KAR values.

The distribution of KAR_{cell} values is presented in Fig. 4.7a separately for grades

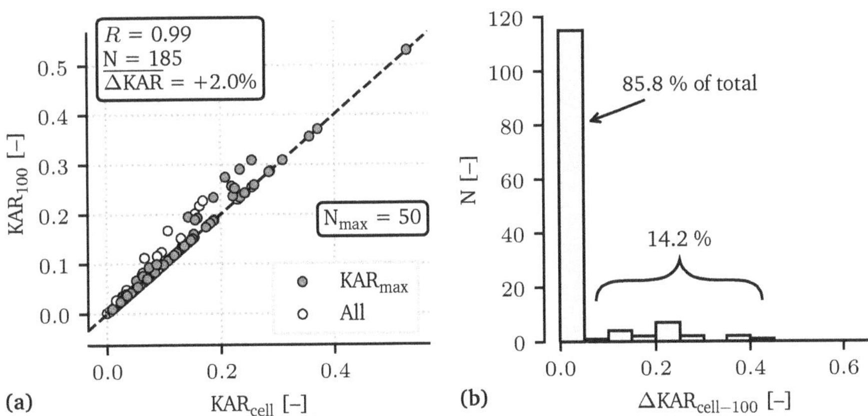

Figure 4.5. Comparison of KAR_{cell} and KAR_{100}. (a) all weak zones; (b) histogram of the residues $\Delta KAR_{cell-100} = (KAR_{cell} - KAR_{100})/KAR_{100}$

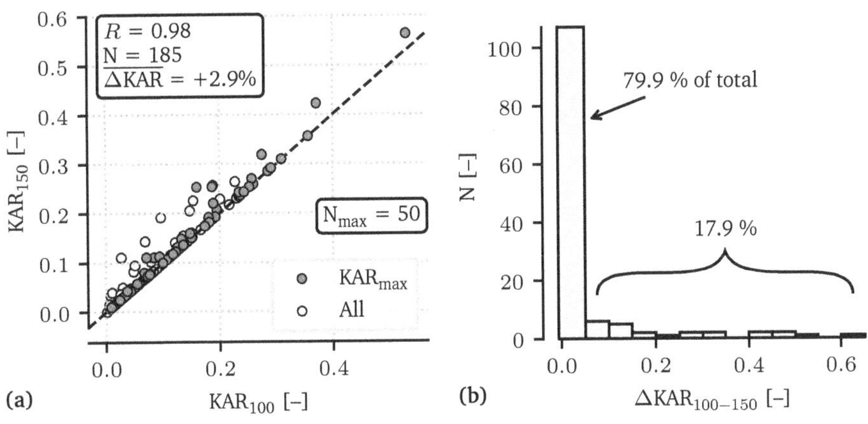

Figure 4.6. Comparison of KAR_{100} and KAR_{150}. (a) all weak zones; (b) histogram of the residues $\Delta KAR_{150-100} = (KAR_{150} - KAR_{100})/KAR_{100}$

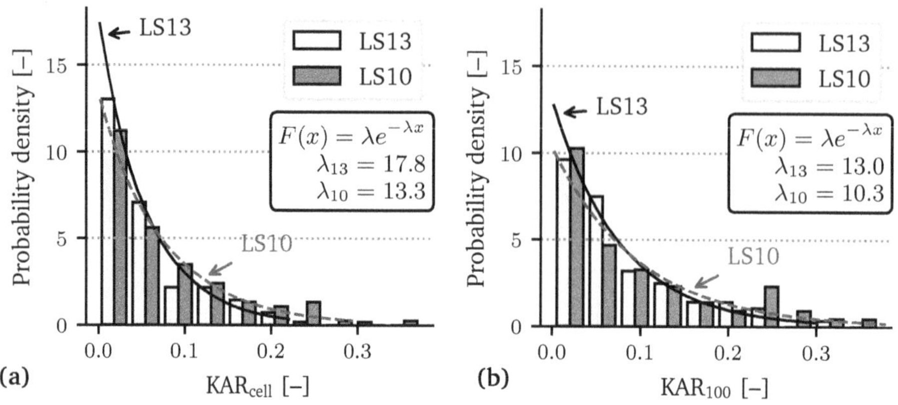

Figure 4.7. Statistical distributions for KAR values, separately for LS10 and LS13 grades. (a) KAR_{cell}; (b) KAR_{100}

LS10 and LS13. Visual inspection of the data suggested an exponential distribution as a good candidate to properly represent the data. The same holds true for the KAR_{100} data presented in Fig. 4.7b. For the case of KAR_{cell}, the distribution presents a larger number of relatively small values, thus decreasing more rapidly compared to the KAR_{100} values. This is more clearly expressed by the λ parameters of the respective fitted exponential distributions: For KAR_{100} the parameters are $\lambda_{10} = 10.3$ and $\lambda_{13} = 13.0$, whilst for KAR_{cell} values of $\lambda_{10} = 13.3$ and $\lambda_{13} = 17.8$ hold. Hence the exponential decay is stronger for the KAR_{cell} data.

Relative KAR values within board

The next analysis considers the relative KAR values within a board, i.e. given a KAR_{max} in a board, how does the n-th highest KAR value ($KAR_{max,n}$) relate to KAR_{max}. In Fig. 4.8a and Table 4.1 the ratio $KAR_{max,n}/KAR_{max}$ is presented for the first five $KAR_{max,n}$ values in each board for KAR_{cell}, as usual separately for grades LS10 and LS13. The individual datapoints are also depicted with the circle markers over each bar. Figure 4.8b illustrates the respective results for KAR_{100}.

The KAR_{100} values tend to decrease at a higher rate as compared to KAR_{cell}, which can be better observed in Table 4.1, where mean and standard deviation values for the n-th relative KARs are presented. It can be seen that a considerable difference is present, starting from the third highest KAR, where in average, the relative magnitude of KAR_{100} is 38 %, whilst for KAR_{cell} it is 48 %. In addition, the scatter for the ratios of KAR_{100} tend to be lower, expressed in lower standard deviations.

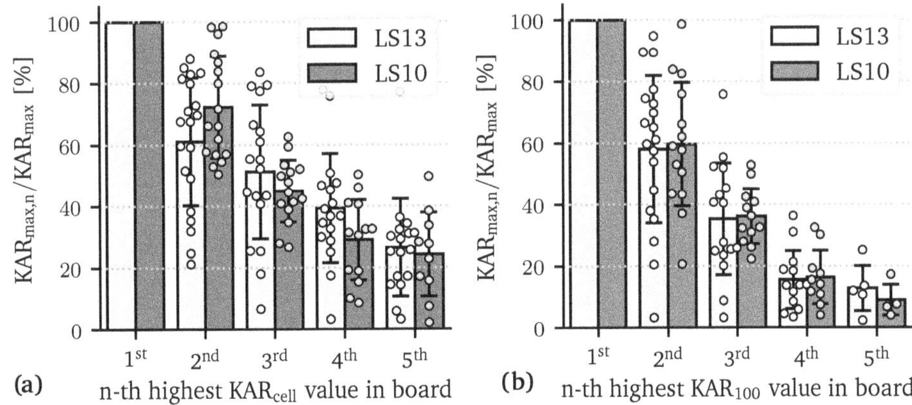

Figure 4.8. Relative KAR values within board for the five n-th largest KARs. (a) Computed with KAR_{cell}; (b) computed with KAR_{100}

Table 4.1. Relative value of the n-th largest KAR within each board as percentage of KAR_{max}

		All [%]	LS10 [%]	LS13 [%]
	Data	μ σ	μ σ	μ σ
Second highest KAR	KAR_{cell}	66.2 ± 20.0	61.0 ± 21.3	72.3 ± 17.1
	KAR_{100}	58.7 ± 22.7	58.0 ± 24.6	59.6 ± 20.8
Third highest KAR	KAR_{cell}	48.4 ± 18.0	51.3 ± 22.4	44.9 ± 10.4
	KAR_{100}	35.7 ± 15.2	35.3 ± 18.8	36.1 ± 9.3
Fourth highest KAR	KAR_{cell}	34.9 ± 17.0	39.4 ± 18.3	29.0 ± 13.6
	KAR_{100}	15.9 ± 9.3	15.5 ± 9.8	16.3 ± 9.1
Fifth highest KAR	KAR_{cell}	25.7 ± 15.4	26.5 ± 16.3	24.3 ± 14.4
	KAR_{100}	11.0 ± 7.1	12.7 ± 8.2	8.9 ± 5.8

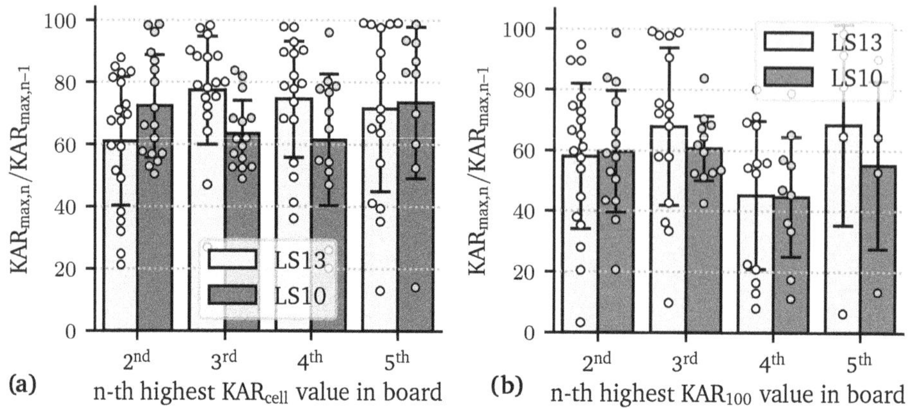

Figure 4.9. N-th largest KAR ($KAR_{max,n}$) relative to the previous largest KAR ($KAR_{max,n-1}$): (a) Computed with KAR_{cell}; (b) computed with KAR_{100}.

In general, the n-th largest relative KAR value within a board follows roughly a geometric progression. This can be corroborated by inspecting the ratios between each KAR value with the previous largest KAR value within a board. This is shown in Figs. 4.9a and 4.9b for both analyzed grades. It can be seen that the ratio oscillates around 70 % and 60 % for KAR_{cell} and KAR_{100}, respectively (see Table 4.2). This means that a geometric progression of the form

$$KAR_n = r^n \tag{4.3}$$

with $r = 0.7$ can represent the evolution of the n-th largest KAR_{cell} value in a reasonable manner. In an analogous manner, a value $r = 0.6$ represents the n-th largest KAR_{100} value.

Maximum KAR value per board

The final step of this section consists on the analysis of the KAR_{max} values, exclusively. Owed to the rather small number of boards, a separate analysis for both grades LS10 and LS13 is not possible. Hence, the boards belonging to the LS10 and LS13 grades are analyzed together. The Gamma distribution was chosen to represent these data based on the findings of Isaksson (1999) and Fink (2014). Figure 4.10 shows the probability density of the $KAR_{100,max}$ values along with a fitted Gamma distribution and its respective shape (α) and rate (β) parameters.

Table 4.2. Relative value of the n-th largest KAR with respect to the previous largest KAR ($KAR_{max,n-1}$)

	Data	All [%] μ	All [%] σ	LS10 [%] μ	LS10 [%] σ	LS13 [%] μ	LS13 [%] σ
Second highest KAR	KAR_{cell}	66.2	± 20.0	61.0	± 21.3	72.3	± 17.1
	KAR_{100}	58.7	± 22.7	58.0	± 24.6	59.6	± 20.8
Third highest KAR	KAR_{cell}	71.1	± 16.6	77.4	± 17.9	63.4	± 11.1
	KAR_{100}	64.8	± 21.5	67.8	± 26.8	60.7	± 11.1
Fourth highest KAR	KAR_{cell}	68.9	± 21.2	74.6	± 19.3	61.5	± 22.0
	KAR_{100}	45.0	± 23.0	45.3	± 25.4	44.7	± 20.8
Fifth highest KAR	KAR_{cell}	72.3	± 26.3	71.6	± 27.4	73.5	± 25.7
	KAR_{100}	62.6	± 33.4	68.4	± 37.0	55.2	± 31.9

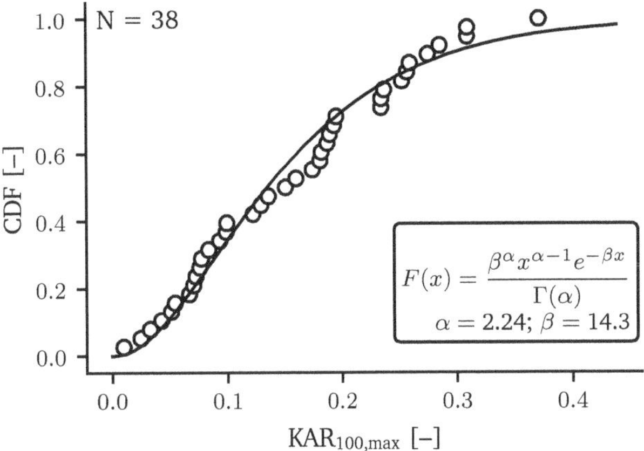

Figure 4.10. Cumulative probability distribution for $KAR_{100,max}$ values of combined grades LS10 and LS13

4.3 Length of clear wood segments

In this section an analysis of the segment lengths of clear wood is presented. The intention of this is to obtain the necessary statistical information regarding the inter-knot distances, which is needed for the simulation of knots (KAR values) along virtually generated boards.

4.3.1 Definition of clear wood length

The clear wood length (ℓ_{cw}), or inter-knot distance, is defined in Fig. 4.4 as the distance between the center points of two consecutive weak zones, according to KAR_{100} data. The distance between the start (end) and the first (last) WZ are considered here as clear wood segments as well. For the latter case, 50 mm are added (half the length of the window) for consistency reasons with the other measurements. If no knots are present on the length of board, then ℓ_{cw} is equal to the length of the board. This definition is comparable to the one used by Fink (2014).

For the case that the KAR_{cell} data is used, then the distance ℓ_{cw} is defined as the distance between the center of two consecutive WZs, or the distance between the start (end) of the board and the first (last) WZ plus 50 mm (for the same reasons as above). Knot-free boards are also considered in the analysis, as explained in the case of KAR_{100} data.

4.3.2 Clear wood length analysis

The clear wood length was computed for each board considering different minimum KAR values as thresholds (KAR_{thres}). This approach is meant to give a more in-depth insight into the distribution of distances between WZs of KARs with different magnitudes. Such information potentially allows for a more accurate simulation of the positions of WZs along boards. The used KAR_{thres} values are 0.0, 0.05, 0.10, 0.15 and 0.20, which are used to define the groups of different ℓ_{cw} specified in Table 4.3. The analysis was performed with both KAR_{cell} and KAR_{100} data to illustrate the effect of the KAR definition on the ℓ_{cw} distributions.

Figure 4.11a presents the ℓ_{cw} results separately for LS10 and LS13 grades for the KAR_{cell} data. For the case that all WZs are considered ($KAR_{thres} = 0$), the average distance ℓ_{cw} is about 320 mm for both grades, with a standard deviation, σ of 330 mm (see Tab. 4.4 for more details). As the threshold KAR_{thres} is increased, ℓ_{cw} also grows. An interpretation for this is that larger KAR values are in general more amply spaced between each other. There are clear differences between the grades LS10 and LS13.

For small KAR_{thres} both grades show similar values for ℓ_{cw}. However, for higher KAR_{thres} values, larger clear wood distances are observed for LS10 as compared to LS13 boards. This difference is most probably bound to the classification criteria used (DIN 4074-5, 2008), since the presence of higher KARs (LS10) could be related to more WZs in general. Additionally, the observed difference might be

Table 4.3. Definition of the KAR values considered for each clear wood length group $\ell_{cw,k}$

ℓ_{cw} group:	$\ell_{cw,0.0}$	$\ell_{cw,0.05}$	$\ell_{cw,0.10}$	$\ell_{cw,0.15}$	$\ell_{cw,0.20}$
Definition:	$0 < KAR$	$0.05 < KAR$	$0.1 < KAR$	$0.15 < KAR$	$0.2 < KAR$

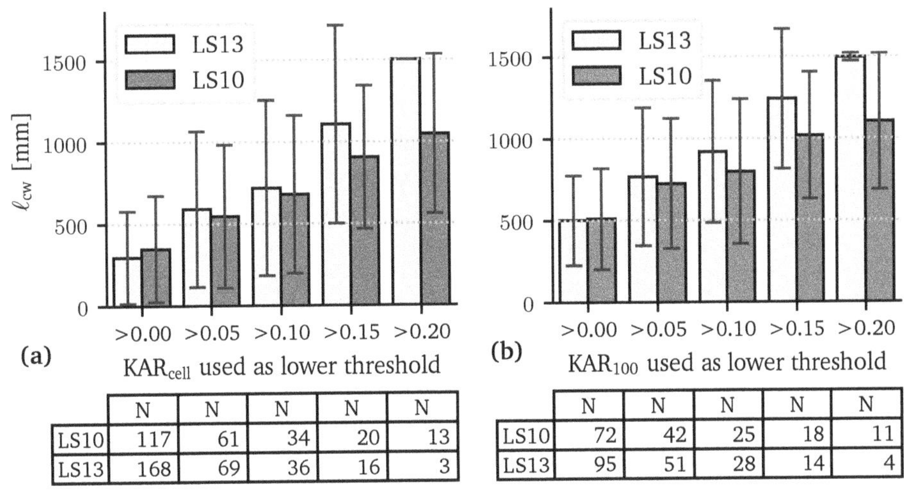

(a)

	N	N	N	N	N
LS10	117	61	34	20	13
LS13	168	69	36	16	3

(b)

	N	N	N	N	N
LS10	72	42	25	18	11
LS13	95	51	28	14	4

Figure 4.11. Clear wood length (ℓ_{cw}) for different KAR thresholds. (a) computed with individual cells; (b) computed with knot clusters identified with the KAR_{100} values. Note: The lower tables in both figures give the number of clear wood segments used in each bar.

linked to the presence of a higher number of knot-free boards among the LS13 grade, increasing the average ℓ_{cw}.

Figure 4.11b shows the ℓ_{cw} results obtained using the KAR_{100} data. The general trend is the same as for KAR_{cell}, except for the slightly higher values of ℓ_{cw} for lower KAR_{thres} (520 mm for $KAR_{thres} = 0$, compared to 320 mm for the KAR_{cell} data; see Tab. 4.4 for detailed results). This is to be expected, as KAR_{cell} considers each cell with $KAR > KAR_{thres}$ individually, whilst the definition of WZ using KAR_{100} tends to *merge* adjacent cells with $KAR > 0$ into one larger weak zone, thus preventing the very small distances associated to such cases.

4.4 Final remarks

This chapter presented an analysis of different knot-related parameters of the studied oak boards. Specifically, the distribution of KAR values, maximum and relative KAR values per board, as well as clear wood length were determined and discussed. All these parameters could be combined to produce a simulation model for KAR values within board, e.g. similar as done by Blaß et al. (2005) or Fink (2014). However, in this thesis, the KAR values are mainly used for the analysis of the variation of MOE along boards, described in the next chapter. Although the development of a simulation model for KAR was one of the original objectives of

Table 4.4. Clear wood length (ℓ_{cw}) for individual cells (KAR_{cell}) and with the knot clusters identified with the KAR_{100} values

		All [mm]	LS10 [mm]	LS13 [mm]
	Data	μ σ	μ σ	μ σ
$\ell_{cw,0.0}$	KAR_{cell}	318 ± 300	349 ± 323	296 ± 281
	KAR_{100}	506 ± 290	510 ± 309	502 ± 275
$\ell_{cw,0.05}$	KAR_{cell}	569 ± 459	546 ± 437	590 ± 476
	KAR_{100}	749 ± 412	726 ± 397	768 ± 422
$\ell_{cw,0.10}$	KAR_{cell}	700 ± 511	679 ± 483	719 ± 535
	KAR_{100}	862 ± 442	798 ± 442	919 ± 435
$\ell_{cw,0.15}$	KAR_{cell}	994 ± 529	905 ± 438	1106 ± 606
	KAR_{100}	1115 ± 420	1017 ± 387	1242 ± 427
$\ell_{cw,0.20}$	KAR_{cell}	1131 ± 474	1046 ± 488	1500 ± —
	KAR_{100}	1208 ± 396	1103 ± 415	1495 ± 24

this analysis, further insights gained during the course of this work shifted the focus to an alternative approach, where the explicit consideration of knots is not needed.

CHAPTER 5

VARIATION OF MECHANICAL PROPERTIES IN OAK BOARDS

5.1 General

The study of the variation of mechanical properties between and within boards is an inevitable step in any attempt at developing a glulam strength model. A deep understanding of the variation of MOE, tensile strength and density of the boards, and the relationships between them, enables the development of a more accurate material simulation model for the boards. Once included in a glulam strength model these material models, simulates in a realistic manner the distribution of strength and stiffness throughout the simulated beams. The correct representation of the lamination effect depends in great part on this.

The main objective of this chapter is to characterize and quantify the degree of variation of MOE, tensile strength and density between and within oak (*Q. petraea*, *Q. robur*) boards, and to analyze the relationships that might exist between them. In particular, the variation of MOE will be analyzed within the frame of autoregressive models, giving a notion of the spatial correlation within a board. Then, the localized (cell-wise) tensile strength will be studied by means of censored analysis. All the data presented in this chapter are available in Tapia and Aicher (2020) and Tapia and Aicher (2021).

5.2 Description of the experimental campaign

The 47 boards form the Dataset A were used for a comprehensive experimental campaign with the objective of analyzing the variation of different material properties between and within boards. Objects of this investigation were the MOE, tensile strength and density. In the following, the experimental procedure used for the determination of each of these variables is presented.

5.2.1 Variation of modulus of elasticity along board

A central aspect of the experimental campaign concerned the measurement of localized values of MOE along the main axis of the oak boards. To this end, the central region of each board, with a total length of 1500 mm ($\ell_{E,\text{glob}}$), was considered (see Fig. 5.1). This region was subdivided into 15 consecutive cells (segments) with equal lengths of $\ell_{i,\text{cell}} = 100\,\text{mm}$ (same cells as used for the analysis of knot variables).

For each cell the axial elongation was measured by means of a specific extensometer, illustrated in Figs. 5.2a,b. The device consists of two U-shaped steel frames, attached to the board by means of special screws. The frames are equipped with two linear variable displacement transducers (LVDTs), placed at diagonally opposite sides of the cross-section of the board (see Figs. 5.2a,b). The used LVDTs correspond to the model D6/0500ARA-L25, company RDP Electronics, Great Britain. For the experimental determination of global MOE, $E_{t,0,\text{glob}} = E^{\text{test}}_{t,0,\text{glob}}$, the gauge length was $\ell_{E,\text{glob}}$, spanning over the 15 cells of 100 mm in length.

The tensile tests were performed with a servo-hydraulic universal testing machine (max. load 700 kN), equipped with a pair of special hydraulic-controlled grips, which enable a rather smooth stress transmission from the grips into the board. The boards were placed in the testing machine as shown schematically in Fig. 5.1. The

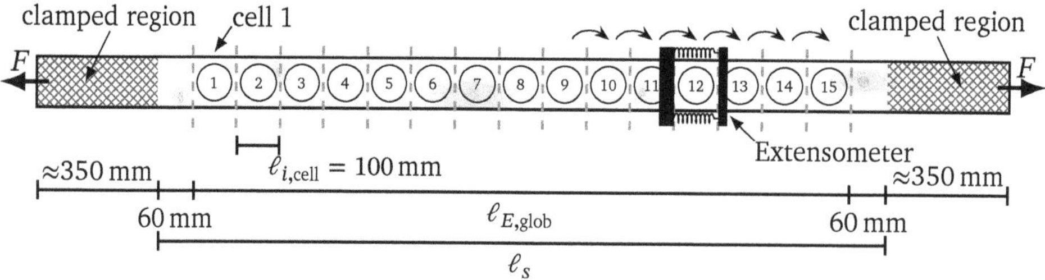

Figure 5.1. Experimental setup used for the determination of the variation of modulus of elasticity along the length of the boards

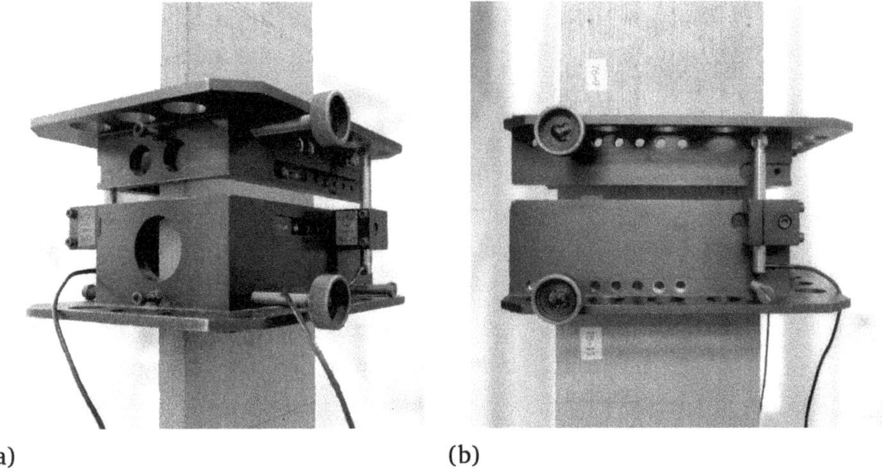

(a) (b)

Figure 5.2. Extensometer used for the measurement of the deformations in the 100 mm long cells. (a) rear-side view; (b) lateral view. The used LVDTs can be seen in both figures.

free length between the grips was throughout $\ell_s = \ell_{E,\text{glob}} + 2 \cdot 60\,\text{mm} = 1620\,\text{mm}$. This resulted in clamping lengths at both ends of the boards of about 350 mm.

The loading protocol of the boards consisted of a series of loading-unloading cycles. In each cycle the axial elongation of one cell of the board was measured within the linear elastic range up to a tensile load of 30 kN, equivalent to a nominal tensile stress of $\sigma_{t,0} = 7.1\,\text{N/mm}^2$. The loading was performed in displacement control, at a stroke rate of 0.04 mm/min, while unloading to zero load in each cycle was done load-controlled at a rate of 5 kN/s. After each loading-unloading cycle, the extensometer was shifted to the next cell and the described sequence of loading, displacement measurement and unloading was repeated for each of the 15 cells within $\ell_{E,\text{glob}}$ (see Fig. 5.1).

In a subsequent step, the global MOE, $E_{t,0,\text{glob}}^{\text{test}}$, was measured over the gauge length $\ell_{E,\text{glob}}$. The objective of the $E_{t,0,\text{glob}}^{\text{test}}$ measurement consisted in both: (i) experimental determination of the global member stiffness and (ii) verification of the overall quality of the MEO measurements at the individual cells. If the local measurements are correct, then the difference between the measured global MOE and the global MOE computed from the local measurements (cell data) should be small in relative terms.

5.2.2 Multiple tensile strength measurements

The measurement of the variation of tensile strength ($f_{t,0}$) along the board presents its own challenges, mainly due to the impossibility of testing consecutive cells in tension until failure due to the needed clamping length. Taking this into consideration, the following approach was chosen to maximize the amount of meaningful information obtained from each board. The boards were tested in tension parallel to the grain direction until failure at a constant piston displacement rate of 0.04 mm/min with a computer-controlled servo-hydraulic testing machine. Both ends of each board were fixed by hydraulically actuated grips, which gradually introduce the clamping pressure along the clamping length ($\approx$ 350 mm), minimizing stress concentrations at the transition of free length to the clamps. Each board's first testing to failure delivered the global tensile strength, $f_{t,0,\mathrm{glob}}$, corresponding to a free length, ℓ_s, of 1620 mm = $9.25 \cdot b$, which conforms closely to the provisions of EN 408 (2012) and EN 384 (2016), where $\ell_s = 9 \cdot b$ is stipulated. As the employed free length is slightly (3 %) longer as specified in the European standards, the obtained global tensile strength values can be regarded being conservative to a very small degree, which means that the obtained strength values tend to be slightly lower (about 1 %) as compared to the standard approach[1].

The two remaining parts of each broken board left after the first global failure test were tested, if possible, in a second and occasionally third and forth tensile test, similar as done by Lam and Varoglu (1991a), then for Spruce-Pine-Fir boards of 6.1 m in length. The location, i.e. the number of the cell responsible for the failure was recorded. The procedure of sampling multiple tensile strength values from one board is illustrated in Fig. 5.3. The outlined procedure was fostered by the predominant occurrence of blunt failures of the tested oak boards, producing two remaining parts separated at the first (global) failure by fracture planes rather perpendicular to the board's length axis (see Figs. 5.14a,b). The free lengths ($\ell_{s,2,i}$) between the grips of these secondary tensile tests were significantly lower as compared to the primary global test. The values for $\ell_{s,2,i}$ were in average ($\pm$ std.) $(3.2 \pm 1.7) \cdot b$ with a minimum length of $0.6 \cdot b$. The reduced lengths of the remnants, however, do not constitute a methodological problem, as the secondary tests are intended to detect the strengths of the stronger parts of the boards in relation to the weakest section in the board, in order to obtain a larger database for local MOE, density and strength correlations.

[1]The mentioned strength decrease results from the fact that there exists a higher probability of more, larger growth-bound defects (within the limits of the respective grading/strength class) in an increased free test-length, as compared to shorter ones. The stated quantification of the tensile strength reduction as a result from the increased free test length is based on the below presented length effect simulation results.

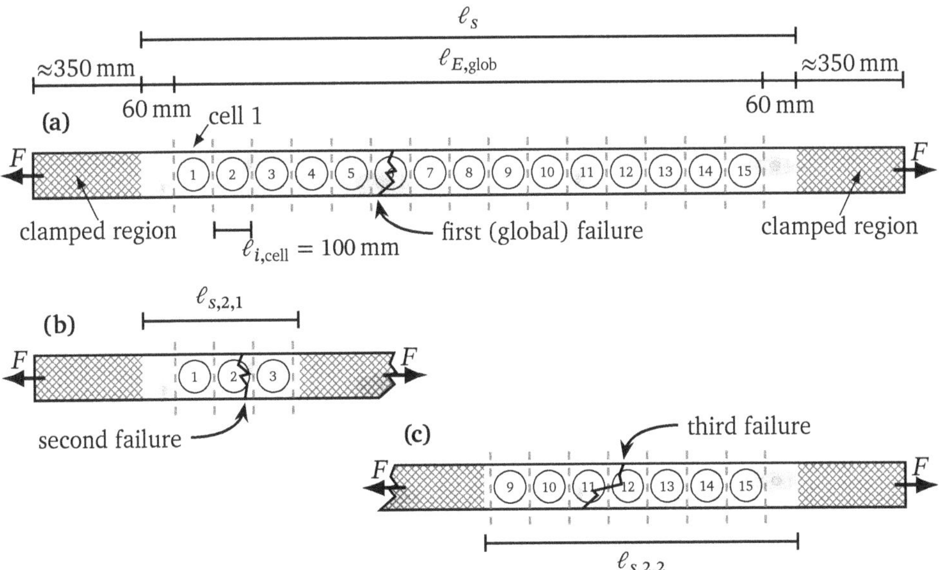

Figure 5.3. Test methodology employed for the gathering of multiple tensile strength values per board

5.2.3 Density measurements along boards

After the tensile tests, the variation of the density along the length of each board was determined. For this purpose, each board was segmented/sawn into its individual cells with a nominal length of 100 mm by means of a circular saw (blade thickness of 4 mm). Clearly, the determination of the density of each of the 15 cells per board was not possible, since the boards were broken in as many as four pieces, depending on the number of secondary tensile tests that each board was subjected to.

In many cases, the cells contained one or two cracks rather parallel to the fiber. These cells were secured by an adhesive tape before cutting. By the described process an average of 11±3 cells was obtained per board for the local cell density measurements. The minimum and maximum number of cells per board suitable for local density measurements was 3 and 15, respectively. The density measurement was performed by weighing and measuring the dimensions of each cell. The necessary determination of the moisture content (MC) for adjustment to the reference density at 12 %MC was done by the oven dry method with two cross-sectional slabs of dimensions 175 × 50 × 25 mm cut from both ends of each board (in the clamped region); the distance from the ends of the boards was throughout 100 mm. The MC of each board was then taken as the average of both measurements.

Table 5.1. Statistics for the global densities normalized to a 12 % ($\rho_{glob,12}$) MC, separated by strength grade

Grade	N	Mean	Std	COV	Min	Max	x_{05}
	Global density, $\rho_{glob,12}$ [kg/m^3]						
Reject	4	725	34	4.6 %	698	783	670
LS7	4	709	69	9.7 %	631	818	597
LS10	17	701	48	6.8 %	634	812	622
LS13	22	691	40	5.8 %	634	762	626
LS10+LS13	39	695	44	6.3 %	634	812	623
All	47	699	47	6.7 %	631	818	623

Table 5.2. Statistics for the local densities normalized to a 12 % ($\rho_{cell,12}$) MC, separated by strength grade. Computed for all the cells of all the boards together

Grade	N	Mean	Std	COV	Min	Max	x_{05}
	Local density, $\rho_{cell,12}$ [kg/m^3]						
Reject	53	712	33	4.7 %	640	812	657
LS7	34	715	75	10.4 %	606	825	592
LS10	196	704	52	7.4 %	609	843	618
LS13	244	686	46	6.8 %	595	826	609
LS10+LS13	440	694	50	7.2 %	595	843	612
All	527	697	51	7.3 %	595	843	613

5.3 Density variation within boards

The statistical evaluation of the global and local density measurements adjusted at 12 % MC is presented in Tables 5.1 and 5.2, respectively. It can be seen that the mean and minimum values of the global density is very similar for grades LS7, LS10 and LS13; the density scatter for grades LS10 and LS13, denoted by low COVs of 7.4 % and 6.8 %, respectively, show no relevant difference. The results indicate that the mean density of the boards is most likely a bad predictor for the strength grade, as there exists no positive correlation with increasing grade.

With regard to the representativity of the investigated oak material in terms of density, a very good agreement with the results of two comprehensive investigations (Glos and Lederer, 2000; Faydi et al., 2017) on oak from different growth regions can be stated. The first mentioned study on bending of 340 unseasoned oak scantlings (growth region south-west Germany, forest district Herrenberg), cut primarily from second and third length segments of logs harvested in thinning operations, was conducted with different cross-sections from 40 × 80 mm to

Table 5.3. Statistics for the COV of local densities for each board

	COV of density of individual boards [%]					
	N	Mean	Std	COV	Min	Max
Grade						
Reject	4	3.6	0.6	17	2.7	4.4
LS7	4	2.9	1.5	51	0.8	4.9
LS10	17	2.5	1.1	46	1.3	5.3
LS13	22	3.0	1.5	52	0.6	8.4
LS10+LS13	39	2.7	1.4	51	0.6	8.4
All	47	2.8	1.4	49	0.6	8.4

60 × 180 mm. The second referenced work on French oak wood (source Bourgogne Franche Comté) focusing on vibrational MOE prediction, comprised 160 boards with cross-sections of about 24 mm × (80 to 170 mm). In both investigations, dealing similarly with a mixture of strength classes D18 to D30 and reject, too, the average gross density, $\rho_{12,glob}$, was about 710 kg/m^3, which is very close to the material regarded here.

The local densities show a very similar picture with regard to its indifference to strength grades. This is emphasized by the fact that the extreme values of local densities reveal almost no difference between the respective grades, disregarding the rejected boards. Regarding the absolute differences between minimum and maximum densities, these are more pronounced on the local level. There, in average, for the grades LS7, LS10 and LS13, a value $\Delta\rho_{min/max} = 226$ kg/m^3 was found. On the global level, the density span is 30 % smaller.

The mean local density variation within a board, characterized by an average COV of 2.8 %, is very small and similar throughout all strength grades (see Table 5.3). However, the spread of the local density variation (COV and extreme values) is somewhat larger for the strength grade LS13 (see Table 5.3). This is reasonable, as the difference between the clear wood and knot areas is more pronounced at higher strength grades.

The very low density scatter between boards and along individual boards matches the results for the global and local densities of beech wood laminations, as presented by Blaß et al. (2005). According to their findings, the density variation along the board length, measured for 20 boards with consecutive board segments with 200 mm lengths, was so small—specified exclusively graphically and ranging between 40 and 80 kg/m^3 around the respective mean board density—that the density was then considered to be constant along the board length in all simulations. Similar findings and conclusions regarding the simulation in a glulam model have been presented by Colling (1990a) for the case of spruce laminations.

5.4 Modulus of elasticity

5.4.1 Processing of measured data

For the evaluation of local and global MOEs the average of both diagonally opposite LVDT measurements was taken. To compute the MOEs, the data-points were considered starting at 10 kN, in order to avoid possible non-linearities produced by slip movements of the boards in the clamping areas at begin of the loading regime. The analysis of the data showed that all measurements between the loads of 10 and 30 kN, respectively, were within the linear range, which was proven by computed squared correlation coefficients, R^2, almost equal to unity.

In order to verify the correctness of both local and global MOE measurements, the global MOE can be derived analytically, $E_{t,0,\mathrm{glob}}^{\mathrm{cell}}$, from all locally measured MOEs within $\ell_{E,\mathrm{glob}}$, by the well-known principle of springs arranged in series as

$$E_{t,0,\mathrm{glob}}^{\mathrm{cell}} = \ell_{E,\mathrm{glob}} \left(\sum_{i}^{N} \frac{\ell_{i,\mathrm{cell}}}{E_{i,\mathrm{cell}}} \right)^{-1} , \qquad (5.1)$$

where $\ell_{E,\mathrm{glob}}$ is the total length spanning the N cells, $\ell_{i,\mathrm{cell}}$ is the length of each individual cell, and $E_{i,\mathrm{cell}}$ is the MOE measured in each cell. In the given case $\ell_{i,\mathrm{cell}} = \mathrm{const.} = \ell_{\mathrm{cell}}$, and $\ell_{E,\mathrm{glob}} = N \cdot \ell_{\mathrm{cell}}$, which simplifies Eq. (5.1) to

$$E_{t,0,\mathrm{glob}}^{\mathrm{cell}} = N \cdot \left(\sum_{i}^{N} \frac{1}{E_{i,\mathrm{cell}}} \right)^{-1} . \qquad (5.2)$$

5.4.2 Global and local moduli of elasticity

Tables 5.4 and 5.5 present a condensed statistical evaluation of all global and local moduli of elasticity (MOE) results, separately for the three visual strength grades, the combined LS10+LS13 sample and for the entire sample. Regarding the mean values of the global MOEs, $E_{t,0,\mathrm{glob}}^{\mathrm{mean}}$, almost no difference can be stated between the LS10 and LS13 grades. The normalized scatter, i.e. the coefficient of variation (COV), is very similar for grades LS10 and LS13 (COV not discussed for LS7 due to the low number of specimens). The minimum and maximum values of $E_{t,0,\mathrm{glob}}^{\mathrm{mean}}$ are slightly (3 % and 4 %) higher for LS13. The mean and extreme values of LS7, irrespective of the very low number of specimens in the sample, are throughout lower, being in line with the grading provisions.

Table 5.4. Global MOEs, $E_{t,0,\mathrm{glob}}$, corrected to MC = 12 % separated by grade

	Global MOE $E_{t,0,\mathrm{glob}}$ [GPa]					
	N	Mean	Std	COV	Min	Max
Grade						
Reject	4	10.6	1.3	13 %	9.2	12.3
LS7	4	11.2	2.6	23 %	8.1	15.0
LS10	17	11.9	2.1	18 %	8.8	16.0
LS13	22	11.7	1.9	17 %	9.1	16.6
LS10+LS13	39	11.8	2.0	17 %	8.8	16.6
All	47	11.6	2.1	18 %	8.1	16.6

Table 5.5. Local MOEs, $E_{t,0,\mathrm{cell}}$, corrected to MC = 12.0 % separated by grade

	Local MOE $E_{t,0,\mathrm{cell}}$ [GPa]					
	N	Mean	Std	COV	Min	Max
Grade						
Reject	60	11.6	2.6	22 %	2.2	18.6
LS7	60	11.7	3.1	26 %	2.7	19.0
LS10	255	12.2	2.7	22 %	5.5	20.7
LS13	330	12.2	2.7	22 %	2.8	22.2
LS10+LS13	585	12.2	2.7	22 %	2.8	22.2
All	705	12.1	2.7	22 %	2.2	22.2

The mean value of the local MOEs, $E_{t,0,\mathrm{cell}}^{\mathrm{mean}}$, shows a very good agreement with $E_{t,0,\mathrm{glob}}^{\mathrm{mean}}$, presenting 3 % to 4 % higher values for grades LS10 and LS13, respectively. However, the spread of $E_{0,\mathrm{cell}}$ is now considerably larger as compared to $E_{t,0,\mathrm{glob}}$. This is also highlighted by the far wider range of the extreme values of the $E_{t,0,\mathrm{cell}}$ results, spreading from 2.1 GPa to 21.6 GPa, whereas for $E_{t,0,\mathrm{glob}}$ the range is much more narrow, spreading from 8.8 GPa to 16.2 GPa. The scatter of $E_{t,0,\mathrm{cell}}$ is shown for the entirety of values in Fig. 5.4. It can be seen in Tables 5.4 and 5.5 that the COV of $E_{t,0,\mathrm{cell}}$ for all boards is 22.4 %, whereas for $E_{t,0,\mathrm{glob}}$ a smaller value of 17.6 % is obtained. For all the boards, a very good agreement was found between the measured global MOE ($E_{t,0,\mathrm{glob}}^{\mathrm{test}}$) and the computed global MOE ($E_{t,0,\mathrm{glob}}^{\mathrm{cell}}$), derived from Eqs. (5.1) and (5.2) based on the locally measured MOEs (see Fig. 5.5a). In average the ratio of both results was $E_{t,0,\mathrm{glob}}^{\mathrm{test}}/E_{t,0,\mathrm{glob}}^{\mathrm{cell}} = 1.01 \pm 0.01$.

Figure 5.5b shows the correlation between dynamic MOE, $E_{t,0,\mathrm{dyn}}$, and measured global MOE, $E_{t,0,\mathrm{glob}}^{\mathrm{test}}$. A high correlation can be observed ($R = 0.97$), although—as usual for dynamic MOE measurements—the $E_{t,0,\mathrm{dyn}}$ results tend to overestimate the MOE by 4 %. This can be visually corroborated by comparing the linear regression with a diagonal having a slope equal to unity (dashed line in Fig. 5.5b), and noticing that the majority of the data-points lay above the diagonal

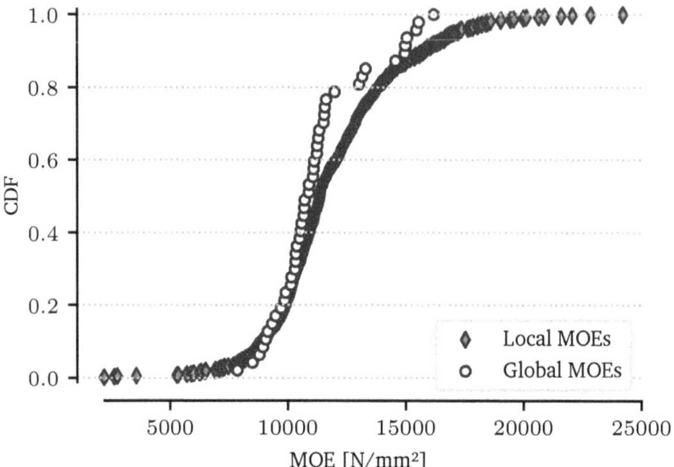

Figure 5.4. Cumulative distribution functions of the global and local MOEs.

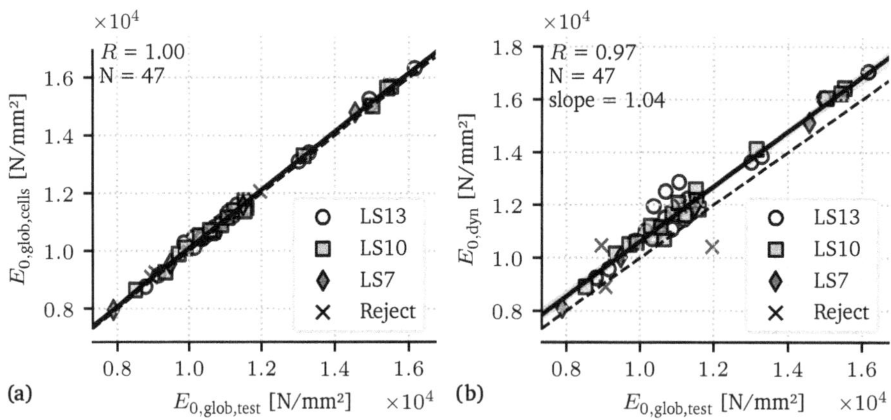

Figure 5.5. Correlation between the different methods used to obtain the global MOE: (a) Correlation between the global MOE computed from the individual local measurements according to Eq. (5.2) and the globally measured MOE. (b) Correlation between the dynamic MOE and MOE measured statically with a gauge length of 1500 mm; The dashed line represents a diagonal with slope = 1.

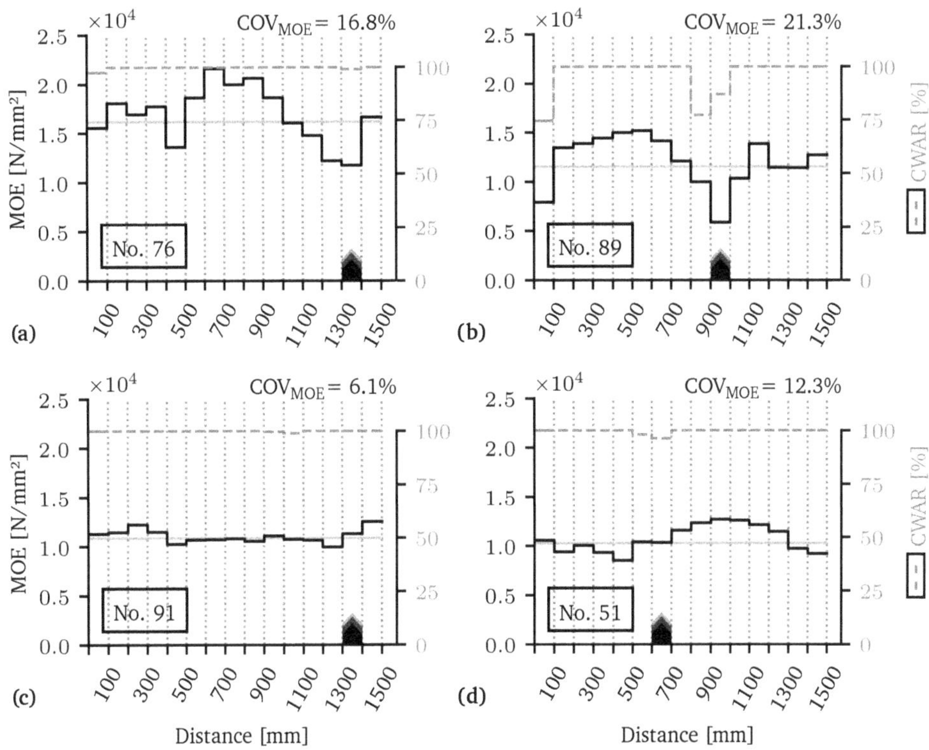

Figure 5.6. Examples of variation of local MOE, $E_{t,0}$, and clear wood area ratios CWAR (dashed line) along the length of different oak boards. The orange, horizontal line denotes the measured global MOE, $E_{glob,test}$. The (cell-)location of the global failure is indicated by the black wedge. (a) Specimen 76 presents a high COV for the MOE which is not explained by a high knotiness; (b) Specimen 89 shows a high COV for the MOE with a very good correlation with CWAR values. (c) Specimen 91 shows a small COV for the MOE with no presence of knots; (d) Specimen 51 shows a moderate COV for the MOE with a small presence of knots;

Table 5.6. Coefficients of variation (COV) of local MOE within individual boards

	COV of $E_{t,0,\mathrm{cell}}$ of individual boards [%]					
	N	Mean	Std	COV	Min	Max
Grade						
Reject	4	18.4	5.8	31.7	11.4	27.6
LS7	4	14.0	5.8	41.3	7.5	21.7
LS10	17	12.9	5.0	38.4	7.0	21.4
LS13	22	11.0	5.5	50.0	3.0	25.7
LS10+LS13	39	11.9	5.4	45.2	3.0	25.7
All	47	12.6	5.7	45.7	3.0	27.6

line.

The MOE fluctuation along the length of the boards is shown in Figs. 5.6a–d for four representative cases. These correspond to the boards No. 76, 89, 91 and 55, belonging to the grades LS13, LS10, LS13 and LS10, respectively. Their corresponding global MOEs ($E_{t,0,\mathrm{glob}}$) are 16.2 GPa, 11.5 GPa, 10.9 GPa and 10.3 GPa, represented by the horizontal line in each figure. Furthermore, the clear wood area ratio (CWAR) of each cell is depicted (dashed line) and the position of the global tensile failure is marked by the black wedge. The results corresponding to the entirety of the tested boards are presented in Appendix A.

Figure 5.6a illustrates a rather high variation of MOE along the board (COV = 16.8 %), without presence of large knots (as indicated by the monotonously high values of CWAR). Thus, the high variation cannot be attributed to the effect of knots in the board, but results most likely from other growth-bound defects, such as the fiber orientation. This contrasts to what is depicted in Fig. 5.6b, where also a rather high MOE variation is observed (COV = 21.3 %). However, this time the variation can be directly correlated to the presence of knots on the board. Another typical MOE-defect feature is presented in Fig. 5.6c, where a very low MOE variation is observed (COV = 6.1 %), accompanied by a low presence of knots. A final example is illustrated in Fig. 5.6d, where a moderate MOE variation (COV = 12.3 %) occurs, resulting mostly from both, knot presence and, presumably, fiber orientation.

The scatter of $E_{t,0,\mathrm{cell}}$ within the boards of the different strength grades is specified in Table 5.6. The mean value of the local MOE scatter is slightly higher for the grade LS10 as compared to LS13. However, the difference in scatter between individual boards, determined by the coefficient of variation of the COV, is pronouncedly more expressed for the grade LS13.

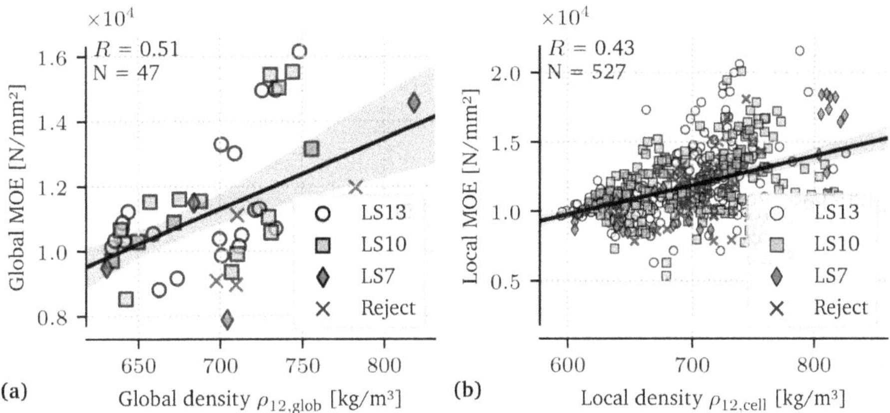

Figure 5.7. Correlations between MOE and density normalized to 12 % moisture content. The data are given separately for the different strength grades LS7, LS10, LS13 and reject. (a) Global level; (b) Local level

5.4.3 Correlation of MOE with density and KAR

Figure 5.7a presents the correlation between the global MOE ($E_{t,0,\text{glob}}$) and global, i.e. total board density ($\rho_{12,\text{glob}}$), denoted by $R = 0.51$; the individual data points are differentiated with regard to the respective board grade.

The reported R-value is substantially lower than the correlations usually obtained for global MOE versus gross density of softwood laminations, where an R-range of 0.7 to 0.9 is typical (e.g. $R = 0.88$ in Colling and Scherberger, 1987). For hardwoods, limited literature data can be found on the relationship between $E_{t,0,\text{glob}}$ and ρ_{glob}. In a recent study on French oak boards an R-value of 0.30 was obtained (Faydi et al., 2017), thus even significantly lower than in this study. For a mixed dataset of beech, oak, and ash laminations, an R-value of 0.40 was reported (Frühwald and Schickhofer, 2005).

Figure 5.7b illustrates the correlation between the local MOE, $E_{t,0,\text{cell}}$, and corresponding local densities, $\rho_{12,\text{cell}}$. In this case, an even lower correlation ($R = 0.43$) compared to the global level was obtained. It should be pointed out that the analyzed dataset is truncated ($N = 527$) regarding the entirety of the cells ($N = 705$), owing to the mentioned fracture-bound impossibility of measuring the density in each cell. The reported MOE-ρ correlation of the board segments is almost twice as high as the respective R-value of 0.23 obtained by Blaß et al. (2005) for a dataset of 330 beech board segments cut from a corresponding number of boards, with an average density $\rho_0 = 680\,\text{kg/m}^3$ and cross-sections in the range of $100 \times 25\,\text{mm}^2$ to $150 \times 35\,\text{mm}^2$. However, contrary to this investigation, in which the local densities are gross values obtained from weighing of the individual 100 mm segments, the density of the beech wood segments was then determined from 20 mm-wide knot-free segment slabs. Although global and, in particular, local MOE-density correlations found in this investigation are substantially higher than those reported for beech laminations, density has to be considered as a rather poor predictor for the MOE within and between oak boards for the material addressed in this study (see Section 5.6). A generalization of this statement to oak species (*Q. robur*, *Q. petraea*) would, however, require a significantly larger database.

The relationship of the global MOE with the grade-determining KAR_{max} value delivers a very low coefficient of correlation of 0.3. The removal of the two highest KAR_{max} ratios reduces the correlation to $R = 0.16$, which highlights the unsuitability of this relative small-sized database for establishing a reliable global MOE-KAR_{max} relationship. Furthermore, the suitability of such a correlation can be questioned with regard to the theoretically very small impact (e.g. Eqs. (5.1)) of a single weak local cell. The linear relationship between the cell-related KAR value, KAR_{cell}, and $E_{t,0,\mathrm{cell}}$, based on a total of 527 cells, yielded, somewhat surprisingly, a similarly weak correlation coefficient of 0.31, being lower as the above reported R-value for the MOE-density relationship. The obtained low $E_{t,0,\mathrm{cell}}$–KAR_{cell} correlation deviates considerably from the result $R = 0.70$ obtained by Blaß et al. (2005) for this relationship in the beech wood study. Whether the stated marked difference is species-dependent, or/and a result of the specifics of the samples in this and the referenced investigations, is of high interest and has to be investigated further.

5.4.4 Normalization of the measured MOE data

The results analyzed in this chapter show a large variation of $E_{t,0,\mathrm{glob}}$ for the studied boards ($\sigma_{E,\mathrm{glob}} \approx 2000\,\mathrm{N/mm^2}$), some of which can be explained by the difference in density observed in the boards (see above). The rest of the global variability may be explained by other characteristics, such as the relative position of the board in the trunk, average fiber orientation, or the presence of sapwood (Kollmann and Côté, 1968). The cell-wise MOE, $E_{t,0,\mathrm{cell}}$, exhibits an even higher total standard deviation ($\sigma_{\mathrm{tot}} \approx 2600\,\mathrm{N/mm^2}$), which results from the superposition of the inter-board variation, $\sigma_{E,\mathrm{glob}}$, and the local level (intra-board) variation ($\sigma_{E,\mathrm{local}}$), in theory complying with the relationship

$$\sigma_{\mathrm{tot}}^2 = \sigma_{E,\mathrm{glob}}^2 + \sigma_{E,\mathrm{local}}^2, \tag{5.3}$$

for the simplified case where both variables are uncorrelated.

As simple as Eq. (5.3) might be, it constitutes the basis for most of the known models to simulate MOE values along boards (see e.g. Ehlbeck and Colling (1987), Blaß et al. (2005), and Fink (2014)). It justifies the hierarchical approach, where firstly a global value for the board is generated based on $\sigma_{E,\text{glob}}$, and then the local variation is considered (based on $\sigma_{E,\text{local}}$). This concept is adopted in the present work too, however, the methodology used differs from that of previous models.

In the following, the characteristics of the local MOE variation are analyzed. For this, a normalization approach will be presented, where the measured localized MOE values of all boards are brought to a common level, allowing for the comparison of the intra-board variation between the different boards.

The finding of a suited normalization method requires a theoretical understanding of the relevant variables that directly affect the measured $E_{t,0,\text{cell}}$. For example, it seems reasonable to assume that the within-board MOE variation is *mostly* a consequence of the fiber orientation throughout the board. This has been proven to produce very accurate results when comparing measured local bending stiffnesses with values computed on the basis of laser-scanned fiber orientations for Norway spruce (Hu et al., 2018). The same method was applied to French oak by Olsson et al. (2018), however no local bending measurements were available. This assumption implies that higher relative values of $E_{t,0,\text{cell}}$ should correspond to cells with fiber orientations parallel to the main axis of the board (or very close to it), whilst lower relative $E_{t,0,\text{cell}}$ values should coincide with local deviations of the grain with respect to the main axis of the board. Another relevant variable is, of course, the presence of knots. However, this effect may also be captured by the determination of fiber orientation only (Hu et al., 2018; Olsson et al., 2018). An appropriate normalization could then be related to the MOE values in regions free of knots and with no grain deviation, which could be considered as the base value for the MOE in each board.

However, the stiffest segment of a board might not correspond necessarily to the segment with lowest fiber distortion, as other variables might show some influence here, too (e.g. density variation). In addition, it is possible for the fibers of a board to be consistently deviated throughout the length of the board, resulting in an entire absence of cells with fibers parallel to the main axis. For the present analysis, however, this does not represent a major problem, as the objective is to capture the *effect* of the fiber variation (and other variables) on the MOE, and not the fiber orientation itself. Considering this, and in absence of detailed local grain orientation, the base level is here defined as the average of the three largest $E_{t,0,\text{cell}}$ values in each board ($\bar{E}_{\text{max},3}$), being an arbitrary but reasonable normalization value for this purpose. Thus, the normalization was performed based on $\bar{E}_{\text{max},3}$

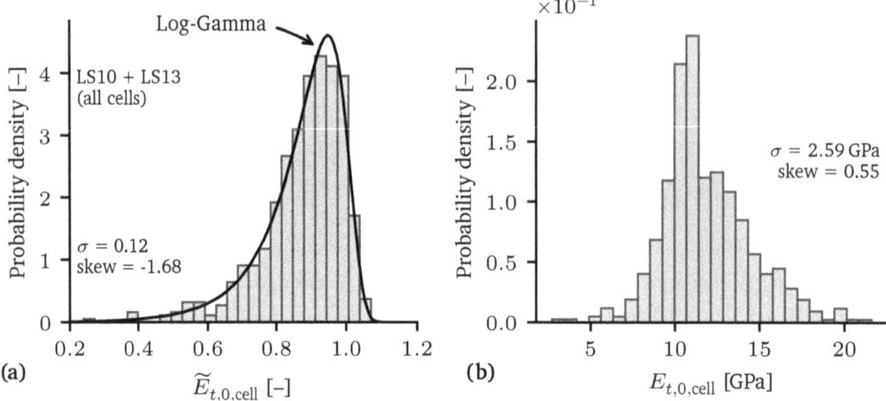

Figure 5.8. Histograms and fitted probability density functions for (a) the normalized MOEs, $\widetilde{E}_{t,0,\mathrm{cell}}$, and (b) measured data, $E_{t,0,\mathrm{cell}}$

according to

$$\widetilde{E}_{t,0,i,j} = \frac{E_{t,0,i,j}}{\bar{E}_{\max,3,i}}, \tag{5.4}$$

where $\widetilde{E}_{t,0,i,j}$ corresponds to the normalized MOE of the cell j of board i, and $E_{t,0,i,j}$ is the corresponding measured local MOE value. If large differences are observed among the two largest $E_{t,0,\mathrm{cell}}$ values in a board (> 10 %), then only the maximum value is used as $E_{\max,3}$.

A histogram of the normalized $E_{t,0,\mathrm{cell}}$ for the boards belonging to grades LS10 and LS13 is presented in Fig. 5.8a. As reference, the non-normalized data is shown in Fig. 5.8b. An important characteristic of the normalized data, $\widetilde{E}_{t,0,\mathrm{cell}}$, is the marked left-skewness (skew=−1.68), which contrasts with the somewhat slightly right-skewed distribution observed prior to the normalization. The skewed $\widetilde{E}_{t,0,\mathrm{cell}}$ reflects the occasional, yet marked relative MOE drop-downs observed within the otherwise rather low variation zones, caused by large grain deviations attributed to the presence of knots, as well as the knots themselves.

The normalized data, $\widetilde{E}_{t,0,\mathrm{cell}}$, was fitted to four candidate distributions: reversed Gumbel ($\mathrm{Gumbel_R}$), Weibull, Log-Gamma and Beta. For both, Weibull and Beta distributions, the location parameter was manually set to zero. Not doing this results in (i) extremely large negative values for the location (order of 10^6) and (ii) equally (but positive) large scale parameter, complicating the direct interpretation of the parameters. Both, $\mathrm{Gumbel_R}$ and Log-Gamma distributions present no

Table 5.7. Fitted parameters for the standardized form of the candidate distributions for (i) all cells and (ii) for clear wood cells

Distribution		loc	scale	shape_1	shape_2	$\log \hat{\mathcal{L}}$	AIC
Gumbel_R	All	0.928	0.0842	–	–	506	−1008
	KAR < 0.05	0.944	0.0717	–	–	504	−1004
Weibull	All	0	0.923	10.1	–	476	−946
	KAR < 0.05	0	0.941	12.7	–	500	−994
Log-Gamma	All	0.988	0.0481	0.437	–	515	−1023
	KAR < 0.05	0.963	0.0618	0.790	–	505	−1003
Beta	All	0	1.07	10.0	2.29	490	−972
	KAR < 0.05	0	1.08	15.5	3.07	503	−998

lower bounds, thus potentially may produce negative values. To prevent this, the distributions were truncated at zero, which does not introduce a large error, given that the cumulative distribution within the disregarded tails is very small.

The estimated parameters for the standardized form of each distribution are presented in Table 5.7, together with the maximum log-likelihood ($\log \hat{\mathcal{L}}$) and the Akaike information criterion (AIC), defined as AIC $= 2k - \log \mathcal{L}$, where k is the number of fitted parameters (Akaike, 1974). Based on this information the Log-Gamma distribution represents the best of the four analyzed models, as it presents the smallest AIC. Having set the location parameter to zero for the Weibull and Beta distributions limits the maximum likelihood obtained for them, affecting the AIC, too. However, the truncation of the lower tail is justified, as it limits the possible lowest MOE values to zero, preventing negative $\widetilde{E}_{t,0,\text{cell}}$ values.

The same analysis was performed considering clear wood sections only, here defined as cells with KAR < 0.05. This is an attempt to separate the effects of the knots on the MOE variability from the *pure* effect of fiber inclination. Specifically, this helps to illustrate the effect of cutting-off board segments containing large knots (similar as in the boards used for the fabrication of the glulam beams of the Dataset B). It is reasonable to expect that the exclusion of large knots would reduce the variation within board, as knot-affected segments should experience in average larger MOE drop-offs compared to knot-free cells, due to the effective reduction of the cross-section and large localized grain deviations.

Figure 5.9a shows the normalized MOE considering clear wood segments only, where a reduction of the spread ($\sigma = 0.09$ compared to $\sigma = 0.12$) can be observed. Additionally, the skew diminishes by about 60 % to a value of −0.98. For this case, the AIC criterion indicates that the Gumbel distribution is the best model, only slightly better than the Log-Gamma model. In fact, the latter is $(-1004 + 1003)/2$

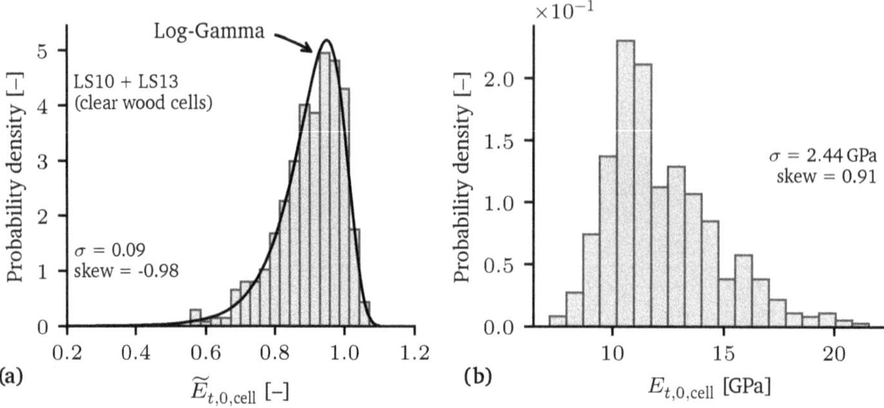

Figure 5.9. Histograms and fitted probability density functions for (a) the normalized MOEs, $\widetilde{E}_{t,0,\text{cell}}$, and (b) measured data, $E_{t,0,\text{cell}}$

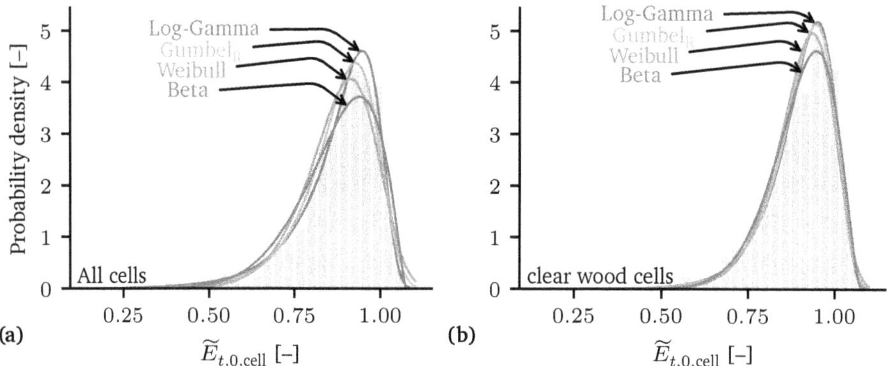

Figure 5.10. Comparison of all fitted distributions for both studied cases: (a) all cells; and (b) cells with KAR ≤ 0.05

≈ 0.606 times as probable as the former to minimize the information loss. This means that both distributions are almost equally appropriate alternatives. All the fitted distributions can be seen together in Figs. 5.10a,b for both analyzed cases, considering (i) all cells and (ii) only cells with KAR ≤ 0.05.

5.4.5 Transformation of normalized MOE to stationary data

One of the most interesting aspects of the measured localized MOE values is the possibility to study the autocorrelation of the $E_{t,0,\text{cell}}$ values along board. The standard procedure to do this involves the application of Eq. (2.15) to the data of *one* board, in order to obtain the sample autocorrelation function (SACF). However, since each board has only 15 cells, the quality of the results would be inadequate. Ideally, the results of all boards should be used together in an aggregated manner, as done e.g. by Taylor and Bender (1991).

A further point concerns the data used to compute the serial correlations, as some theoretical aspects need to be considered. Specifically, computing the SACF directly with the $E_{t,0,\text{cell}}$ values can potentially lead to an overestimation of the autocorrelation, as the pronounced difference of global MOEs between boards has a marked effect on the computed SACF. A solution comes directly from the previous section, where the $E_{t,0,\text{cell}}$ values were brought to a common ground by applying a normalization ($\widetilde{E}_{t,0,\text{cell}}$). Using the $\widetilde{E}_{t,0,\text{cell}}$ values to compute the SACF solves the mentioned issue, as the global variation between boards is suppressed. However, a further issue remains, which has to do with the *stationarity* of the data. By *stationarity* it is meant whether the data is likely to have been produced by a stationary process as described in Section 2.4.3.

To understand the problem, it has to be bared in mind that the main objective of computing the SACF is to calibrate an AR model for $E_{t,0,\text{cell}}$ (see Section 6.3). According to the definitions presented in Section 2.4.3, one condition of an AR model is for its white noise component, ε_i, [see Eq. (2.12)] to be stationary. Recalling the distribution obtained for $\widetilde{E}_{t,0,\text{cell}}$ from the previous section, it is evident that $\widetilde{E}_{t,0,\text{cell}}$ presents a clearly left-skewed distribution, which cannot have originated from a stationary process. A common solution for such cases is to apply a transformation to the data, in order to obtain the needed stationarity. The nature of the transformation is typically determined by inspection of the data.

The transformation chosen here considers the mapping of the distribution describing $\widetilde{E}_{t,0,\text{cell}}$, i.e. ($F(\widetilde{E}_{t,0,\text{cell}})$) into a standard Gaussian distribution $\mathcal{N}(0,1)$. For this, the following equation is used:

$$Z_{t,0,\text{cell}} = \Phi^{-1}\left[F\left(\widetilde{E}_{t,0,\text{cell}}\right)\right], \tag{5.5}$$

where $Z_{t,0,\text{cell}}$ are the *stationary data*, and Φ^{-1} is the inverse cumulative distribution function (CDF, or percent point function) of the $\mathcal{N}(0,1)$. In this manner the stationarity of the data is ensured, and the SACF can be computed.

5.4.6 Serial correlation of localized MOE measurements

The serial correlations were computed using the stationary data, $Z_{t,0,\text{cell}}$, obtained with Eq. (5.5). The data pairs (Z_{i-k}, Z_i) of all boards were used in Eq. (2.15) for the computation of each lag-k. The values Z_{i-k} and Z_i originate from cells of the same board, i.e. the MOE values of different boards are *not* mixed into a single vector to perform the correlation analysis. Placing all the data into a single vector would lead to erroneous results, since cells of different boards would be mixed when computing the serial correlations. Figure 5.11 shows a diagram of the two-step transformation of the local MOE data, $E_{t,0,\text{cell}}$, and its subsequent handling to compute the SACF. For the case of CW cells, a minimum of seven contiguous CW cells were considered in order to obtain meaningful autocorrelations.

A further comment has to be made regarding the described methodology used for the computation of the SACF; this concerns the handling of the data. When computing the serial correlation of a vector $\underline{\boldsymbol{X}}$ (in this case the local MOE along a board) it is obvious that the same results will be obtained regardless of the direction along the board length in which the data are processed. This means that e.g. for the lag-1 it is irrelevant whether the sub-vectors $\underline{\boldsymbol{a}}_1 = [x_1 \dots x_{n-1}]$ and $\underline{\boldsymbol{a}}_2 = [x_2 \dots x_n]$ are taken in one or the other order to compute the correlation (abscissa and ordinate in a graph). However, in the method described here, the stationary MOE values of all the boards are considered together, which raises the question about the *direction* in which the data of each board are concatenated together when computing the correlation. To illustrate this, suppose that now, additionally to the single vector $\underline{\boldsymbol{X}}$, a second vector $\underline{\boldsymbol{Y}}$ is present (representing data from another board), from which also the sub-vectors $\underline{\boldsymbol{b}}_1 = [y_1 \dots y_{n-1}]$ and $\underline{\boldsymbol{b}}_2 = [y_2 \dots y_n]$ are obtained. Then, there are two different possible ways in which they can be concatenated: (i) either $[\underline{\boldsymbol{a}}_1, \underline{\boldsymbol{b}}_1]$ and $[\underline{\boldsymbol{a}}_2, \underline{\boldsymbol{b}}_2]$ or (ii) $[\underline{\boldsymbol{a}}_1, \underline{\boldsymbol{b}}_2]$ and $[\underline{\boldsymbol{a}}_2, \underline{\boldsymbol{b}}_1]$. Both of these options could be used to compute the lag-1 correlation, and since the second halves of the vectors are interchanged, it follows that different values will be obtained.

Knowing this, there are different approaches which can be applied to compute the needed serial correlations. The most trivial one would be to simply take the data of each board in the order they were obtained and concatenate them together to proceed with the computation of the correlations. Another, more complex approach, would be to perform an optimization to find the direction of each board that maximizes the serial correlation. Such an optimization is not trivial, since it corresponds to an integer non-linear problem, which would demand a high computational power even for the low number of boards used. It can be argued, that the difference obtained between these two options might only be significant for a small number of boards, but as the sample size grows larger, the difference should be negligible. For this reason, the first option was used here for the analysis of the serial correlations.

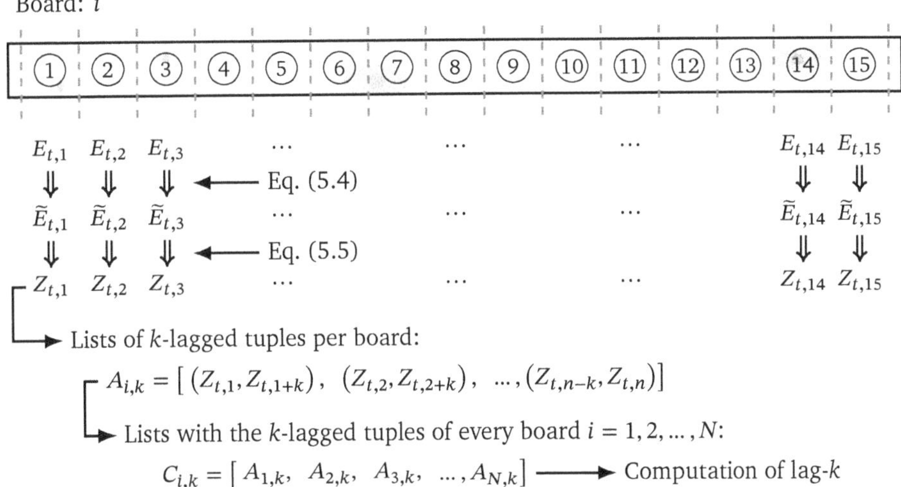

Figure 5.11. Diagram describing the process whereby each *lag-k* autocorrelation is computed

The data used to compute the first five lags are shown in Figs. 5.12a-e for all cells, and in Figs. 5.13a-e for the previously defined clear wood segments. The results obtained for the first five lags are illustrated in Fig. 5.12f and 5.13f for all cells and CW segments, respectively. A steady relative decrease of the correlation level throughout the five lag-correlations can be observed, resembling the characteristic geometric progression of an AR(1) process [see Eq. (2.14)]. The serial correlation results are presented quantitatively in Table 5.8, too. Furthermore, Table 5.8 presents the SACF computed with the original $E_{t,0,\text{cell}}$ values (i.e. without any transformation applied), and additionally with $\widetilde{E}_{t,0,\text{cell}}$, too.

The rather low autocorrelations contrast with the results obtained by Taylor and Bender (1991), where much higher correlations (between 0.95 and 0.89) were found for the first three lags in Douglas-fir lumber. There are many factors that can influence the computed SACF (e.g. species or length of segments), however, the observed difference is most probably explained by the methodology used to compute the SACF. Taylor and Bender did not attempt to normalize the MOEs prior to the regression analysis, as done here. Thus, the computed SACF is probably highly influenced by the variation in global MOE.

This assumption can be tested with the results from Table 5.8, by comparing the SACF obtained with and without the two-step transformation ($E_{t,0,\text{cell}}$ and $Z_{t,0,\text{cell}}$, respectively). Here it can be seen that the SACF values are considerably higher for $E_{t,0,\text{cell}}$, being 64 % and 250 % larger than the results obtained with $Z_{t,0,\text{cell}}$ for the first two lags when all cells are considered (lag-1: from 0.42 to 0.69; lag-2: from 0.17 to 0.59).

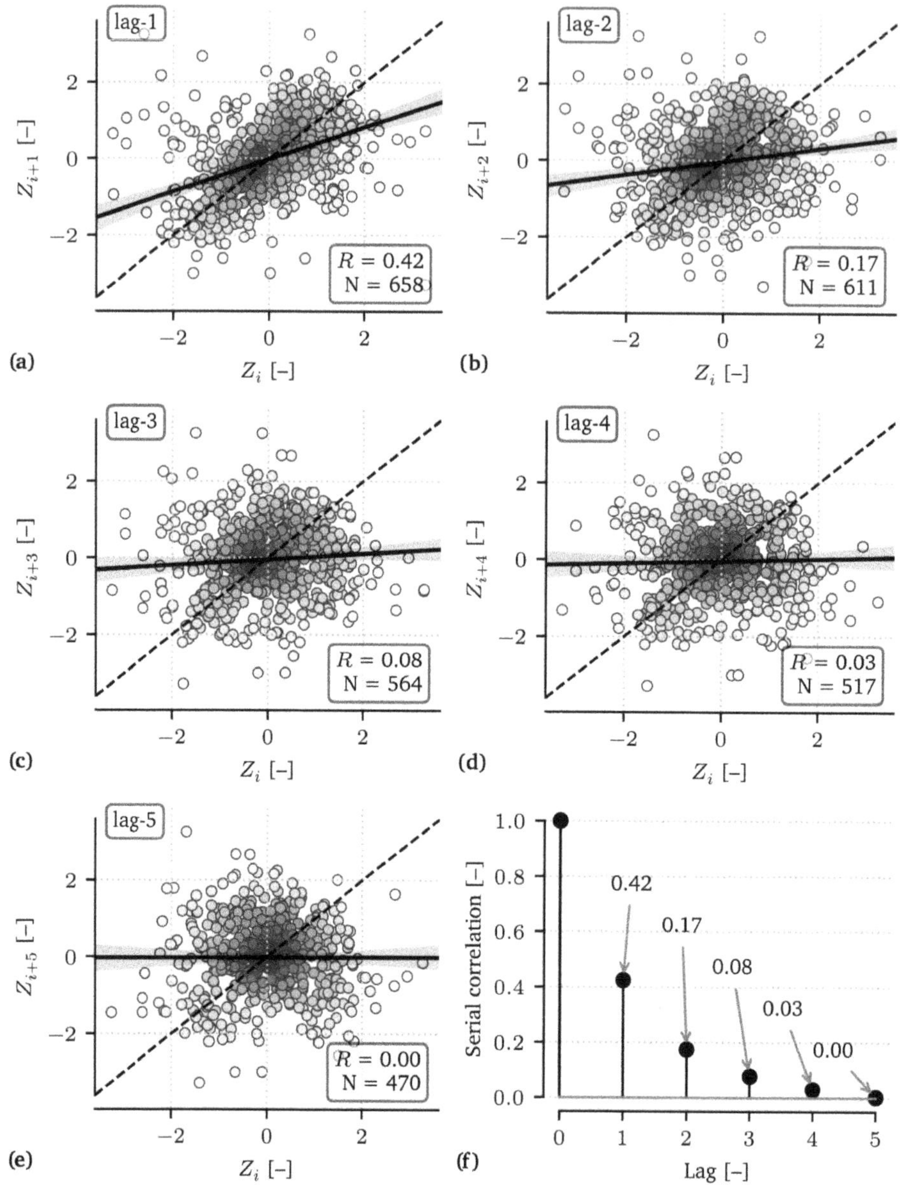

Figure 5.12. Experimentally obtained serial lag-correlations for the local MOE measurements including all cells. (Figs. (a) to (e)) Linear regressions used to compute the first five serial lag-correlations; (f) obtained serial correlations. The dashed line in Figs. (5.12a) through (e) represents a line with slope equal to unity.

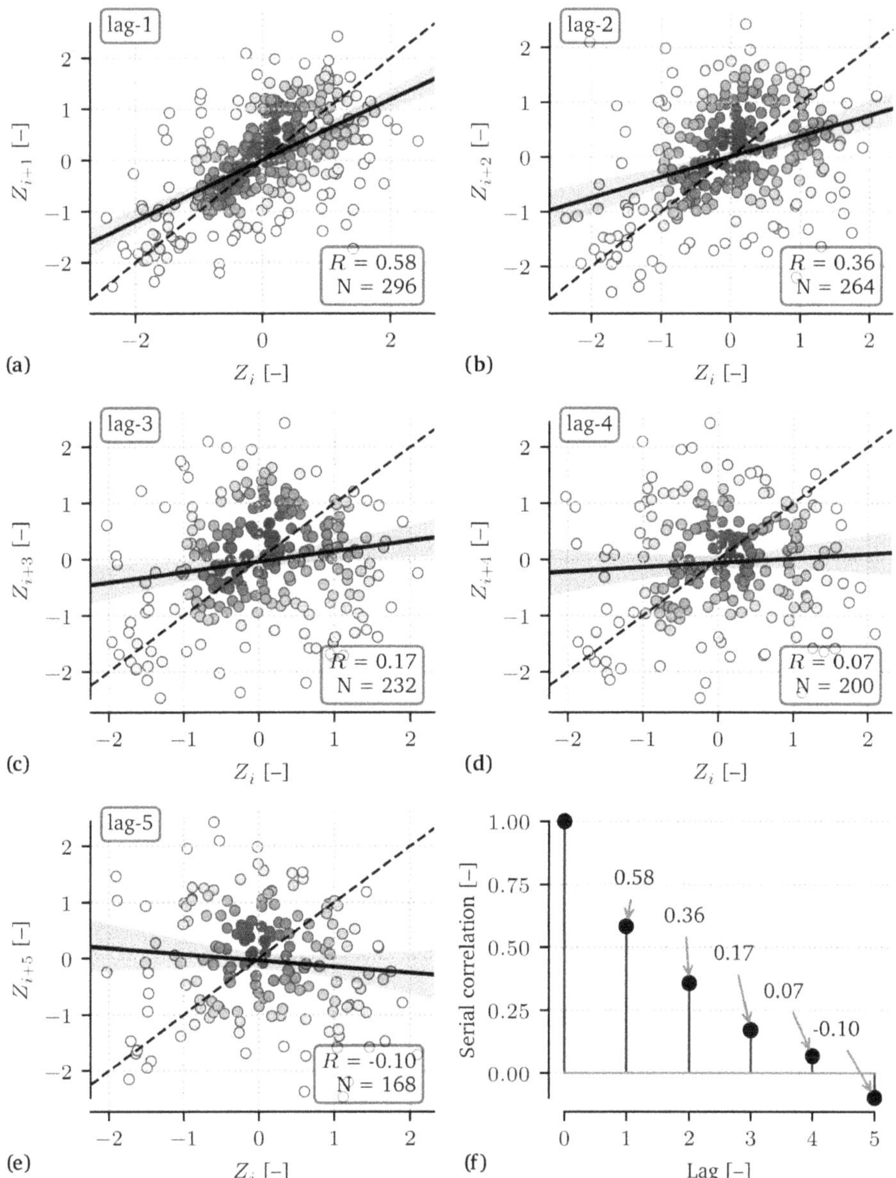

Figure 5.13. Experimentally obtained serial lag-correlations for the local MOE measurements comprising exclusively CW cells. (Figs. (a) to (e)) Linear regressions used to compute the first five serial lag-correlations; (f) obtained serial correlations. The dashed line in Figs. (5.13a) through (e) represents a line with slope equal to unity.

Table 5.8. First five serial correlation values computed for all and for the clear wood segments applying making any additional filtering

		lag-1	lag-2	lag-3	lag-4	lag-5
$E_{t,0,\text{cell}}$	all cells	0.69	0.59	0.59	0.58	0.56
	CW cells	0.89	0.85	0.79	0.75	0.71
$\widetilde{E}_{t,0,\text{cell}}$	all cells	0.40	0.15	0.15	0.13	0.11
	CW cells	0.64	0.44	0.25	0.10	−0.06
$Z_{t,0,\text{cell}}$	all cells	0.42	0.17	0.08	0.03	0.00
	CW cells	0.58	0.36	0.17	0.07	−0.10
ACF AR(1)*	all cells	0.42	0.18	0.08	0.03	0.01
	CW cells	0.58	0.34	0.20	0.12	0.07

* Theoretical autocorrelation function for an AR(1), based on SACF obtained for $Z_{t,0,\text{cell}}$.

When considering only CW cells an increase in the computed autocorrelations is observed, which for the case of $Z_{t,0,\text{cell}}$ is about 53 % (from 0.58 to 0.89). This is a very reasonable result, and supports the intuition that regions with knots, presenting highly localized disturbance in the fibers, exhibit marked lower MOE values, thus reducing the total autocorrelation along the main axis of the board.

The transformation of $\widetilde{E}_{t,0,\text{cell}}$ into $Z_{t,0,\text{cell}}$ has only a minor effect on the computed SACF. For the first lag, and considering all cells, similar values of 0.40 and 0.42 are obtained for $Z_{t,0,\text{cell}}$ and $\widetilde{E}_{t,0,\text{cell}}$, respectively. For the case of CW cells also closely agreeing values of 0.58 and 0.64 (a small decrease of 10 %) are obtained. Finally, and most importantly, it can be seen how closely the SACF follows the theoretical ACF of an AR(1) for the stationary data (shown in Table 5.8 as "ACF AR(1)"). This strongly indicates, that an AR(1) is probably a good model to explain $Z_{t,0,\text{cell}}$. The usefulness of the latter is shown in Section 6.3, where a model for the simulation of $E_{t,0,\text{cell}}$ is presented.

5.5 Tensile strength

The tensile strength obtained from the tests were analyzed at both global and local levels. While the statistical analysis for the global data was performed according to typical procedures (fitting to a statistical distribution and estimating the 5 %-quantile value), different methods were used to assess the local (intra-board) strength variation data. Specifically, survival analysis and order statistics were used to study the within-board tensile strength variation, this being the main topic of this section.

5.5.1 Fracture behavior and global tensile strength

The failure of the boards occurred in all cases in a brittle manner. A highly linear relationship between tensile load and elongation (piston displacement) along the free length ℓ_s was observed. The fracture line extended often rather perpendicular to the board axis and the failure planes showed in many cases a blunt fracture surface (see e.g. Figs. 5.14a,b). Hence, the aspect of the tensile failures is very different from the typical failure modes observed in softwoods and partly in diffuse porous hardwoods (e.g. beech (*F. sylvatica*)), where the crack often follows the fiber direction at the location of the first crack onset, e.g. at a knot. Due to the highly localized failure, the global tensile strengths could be attributed, in general, to one or sometimes two of the cells, for which the local MOE, $E_{t,0,\text{cell}}$, had been previously determined. In most cases (78 %) the tensile failure of the boards started from a knot, and in all other cases from an intersection of inclined fiber orientation with one of the edges of the board.

Table 5.9 presents the statistical evaluation of the global tensile strength values for the different grades and for the structurally relevant grade combination LS10+LS13. The results reveal a small difference between the strength grades LS10 and LS13, where the higher grade, with throughout more restrictive requirements on the visual grading parameters, shows higher strength values on the mean, minimum, maximum and 5%-quantile levels. Here, the 5 %-quantile level was based on a Lognormal distribution at a confidence level of 75 % according to EN 14358 (2016).

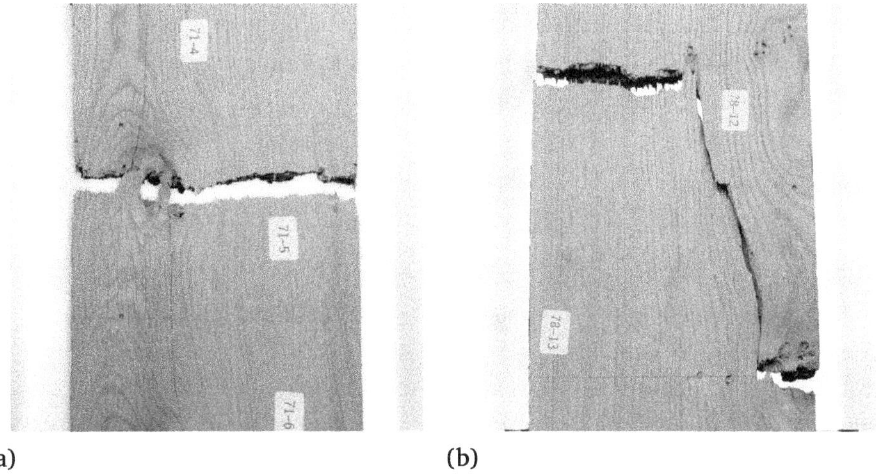

Figure 5.14. Examples of blunt failures obtained in tensile tests. (a) perfectly blunt failure perpendicular to load and fiber direction; (b) mix of blunt failure normal to fiber direction and failure aligned with the inclined fiber direction.

Table 5.9. Global tensile strength results, separately for the different strength grades

		Tensile strength $f_{t,0,\text{glob}}$ [MPa]					
	N	Mean	Std	$f_{t,0,k}$	COV	Min	Max
Grade							
Reject	4	20.8	4.1	13.4	20 %	14.9	25.1
LS7	4	29.6	12.7	7.0	43 %	8.6	40.4
LS10	17	27.4	10.6	12.8	39 %	13.1	49.2
LS13	22	30.5	11.6	13.9	38 %	14.6	55.3
LS10+LS13	39	29.1	11.3	13.4	39 %	13.1	55.3
All	47	28.5	11.2	12.6	39 %	8.6	55.3

The scatter of the strength values of all individual grades, as well as for the combined LS10+LS13 sample, is denoted throughout by a COV of about 40 %, being very high. This leads to rather low characteristic i.e. 5%-quantile values, $f_{t,0,k}$, of 12.8 N/mm^2, 13.9 N/mm^2 and 13.4 N/mm^2 for the grades LS10, SL13 and the grade combination LS10+LS13, respectively.

5.5.2 Local tensile strength

The observed blunt failures of the specimens allowed in many cases (76 %) a further tensile testing of at least one of the remaining parts of the broken board (see Fig. 5.2). A third and fourth tensile test was possible in 43 % and 2 % of the boards, respectively. No evidence for a noticeable weakening of the remnant parts of the boards after the global failure was found.

In a first approach to the analysis, Fig. 5.15a presents all (95) tensile strength values of those boards ($n = 37$) where multiple strength measurements were possible. These data were obtained either in the first loading to fracture, $f_{t,0,\text{glob}}$, or in the secondary loading tests, $f_{t,0,\text{sec}}$. Figure 5.15b presents the same results, now as ratios $f_{t,0,\text{sec}}/f_{t,0,\text{glob}}$. The chosen presentation of the data enables a first assessment of the variability of $f_{t,0,\text{cell}}$ in the studied oak boards.

In both Figs. 5.15a and 5.15b, the boards are ordered from left to right by the assigned hardwood strength grade (LS) and within each LS grade group by ascending strength ratio $f_{t,0,\text{sec}}/f_{t,0,\text{glob}}$ within each board. Table 5.10 contains a statistical evaluation of the results presented graphically in Fig. 5.15b. Very large variations denoted by strength ratios up to almost 6 can be observed. However the mean and standard deviation of the regarded strength ratio of 2.2 ± 1.2 foster the impression that this variation is normally not that extreme. It can be seen that boards belonging to the grade LS13 tend to have lower variation in strength as the LS10 boards (1.93 ± 0.84 % vs. 2.35 ± 1.33 %), which can be attributed

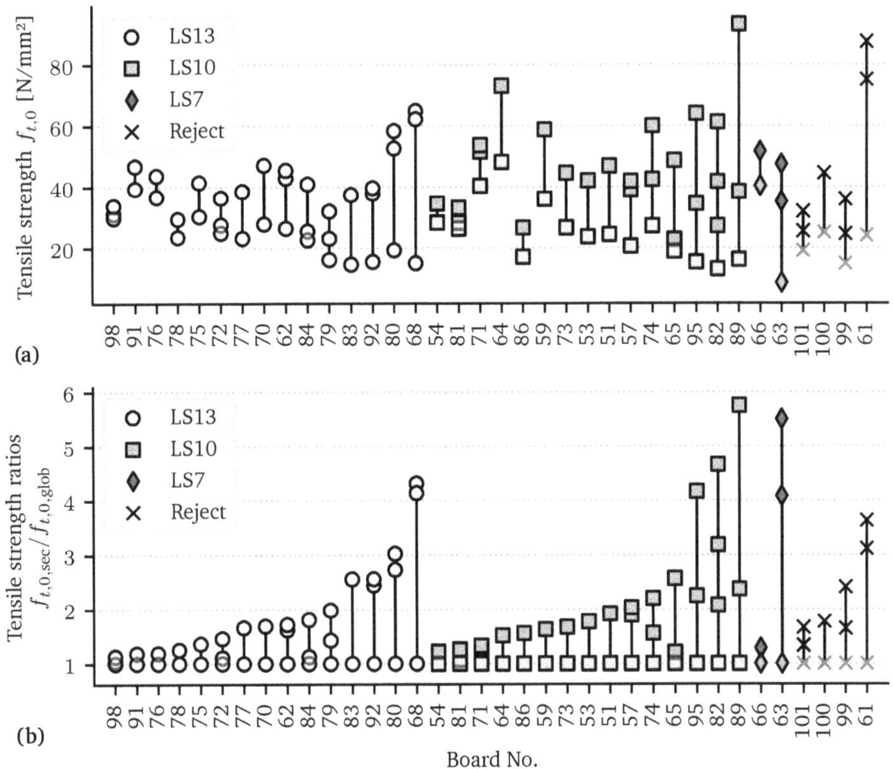

Figure 5.15. Global and secondary tensile strength values of the investigated oak boards of different visual hardwood grades (LS7, LS10 and LS13) according to DIN 4074-5 (2008); (a) absolute values; (b) ratios of secondary strengths vs. global values per board.

to the difference in the allowed knot sizes in each grade. Nevertheless, such an assessment is biased, as the data are highly incomplete, owed to the physical impossibility to test the tensile strength of each cell. As such, the *real* variation should probably be higher. This is analyzed subsequently in more depth.

5.5.3 Relationship between tensile strength and MOE

A regression analysis was performed for the MOE and tensile strength at the global and local levels, for all strength grades together (in Figs. 5.16a,b the respective grades are identified by different symbols). It is apparent that the correlation between $\ln(f_{t,0,\text{glob}})$ and $E_{t,0,\text{glob}}$ is rather weak, denoted by a coefficient of correlation $R = 0.48$. In contrast, the *localized* correlation between $\ln(f_{t,0,\text{cell}})$ vs. $E_{t,0,\text{cell}}$ is significantly higher, characterized by an increased R-value of 0.73. The difference between the two computed correlations is reasonable, as the global tensile strength

Table 5.10. Statistical evaluation of the relative difference of the maximum value of the secondary tensile strengths vs. the global strength of the individual boards.

Grade	Within board strength ratio $f_{t,0,\text{sec}}/f_{t,0,\text{glob}}$					
	N	Mean	Std	COV	Min	Max
Reject	4	2.37	–	–	1.67	3.64
LS7	2	3.39	–	–	1.28	5.50
LS10	15	2.35	1.33	56.8%	1.22	5.75
LS13	15	1.93	0.84	43.6%	1.13	4.32

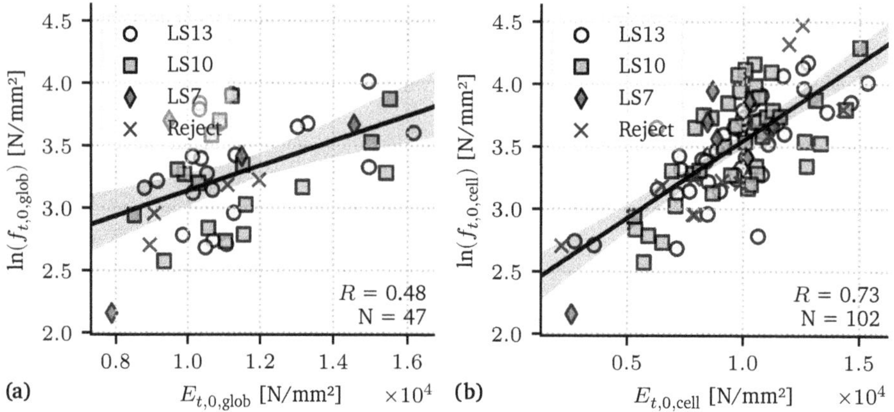

Figure 5.16. Correlations between tensile strength and MOE. (a) global MOE, $E_{t,0,\text{glob}}$, vs. global tensile strength, $f_{t,0}$; (b) local MOE, $E_{t,0,\text{cell}}$, vs. local tensile strength, $f_{t,0,\text{cell}}$.

of a board with the described localized failure is triggered by the weakness of the specific failing cell. Furthermore, the weakest of the serially arranged cells along the full free length, ℓ_S, fails first irrespective of how strong/stiff the board may be at other cells. Hence, it can be expected for the correlation between the local properties $f_{t,0,\text{cell}}$ and $E_{t,0,\text{cell}}$ to be higher as the correlation for the global properties. Similarly, in the secondary tensile tests, performed with the remaining parts after the global failure, the respective weakest cell within the reduced free length fails.

It is tempting to assume that the cells responsible for the global and secondary failures represent the cell with the minimum MOE—or close to it—of all cells within the respective free length, i.e. $E^{\text{fail}}_{t,0,\text{cell}} = E^{\text{min}}_{t,0,\text{cell}}$, as MOE-detrimental variables have probably a negative effect on tensile strength, too. This assumption was checked against the experimental data: the location of $E^{\text{glob-fail}}_{t,0,\text{cell}}$ corresponded with the location of $E^{\text{min}}_{t,0,\text{cell}}$ for 53 % of the tested boards. If the adjacent cells of $E^{\text{min}}_{t,0,\text{cell}}$ are considered too, then 70 % of the global failures occur in the region defined by the position of $E^{\text{min}}_{t,0,\text{cell}}$ and its immediate surroundings. The cases where this does not occur are mostly characterized by boards or board sections with a low $E_{t,0,\text{cell}}$ variation, presumably correlated to a low tensile strength variation. Since many adjacent cells have similar low $E_{t,0,\text{cell}}$ values, the occurrence of the failure at the cell with lowest MOE becomes less likely.

In summary, it can be stated that the weakest cell (responsible for the global failure) is in general associated to the lowest $E_{t,0,\text{cell}}$ of each board, and that $f_{t,0,\text{cell}}$ shows a correlation with $E_{t,0,\text{cell}}$. This is especially true when higher MOE variation is observed within each board. This constitutes some relevant information for the estimation of the intra-board distribution of tensile strength discussed below.

5.5.4 Estimation of the intra-board tensile strength distribution

The presented data give a sense of the amount of variation that can be expected for the tensile strength in the specifically studied oak boards. However, in the presented form these data fail to give an adequate statistical characterization of the variation of $f_{t,0,\text{cell}}$, e.g. a probability distribution. This is owed to the limited amount of data relative to the number of cells defined per board (15 cells). An adequate representation of the variation of $f_{t,0,\text{cell}}$ has a practical importance in the development of a material model for the boards, and is therefore investigated here.

Survival analysis applied to the tensile strength data

The analysis of the variation of tensile strength along board is performed by means of survival analysis (SA). The general methodology for this was presented in Section 2.3.3. In a short summary, survival analysis is based on maximizing the likelihood function, $\mathcal{L}$, of Eq. (2.8).

For the studied oak boards, the censored data correspond to the tensile strength associated to the unbroken cells. For these it is only known that $f_{t,0,i} \geq f_{t,0,k}$, with k being the cell that failed in the corresponding free length (see Fig. 5.17). If the data is considered in this manner, then Eq. (2.8) can be used to study the suitability of different models to describe $f_{t,0,\text{cell}}$.

In this analysis, four different models were fitted to the data. Firstly, there are two censored parametric models: (a) a three parameter Weibull distribution and (b) a Beta distribution; and secondly, two censored regression models with the same base distributions as before: (c) Weibull and (d) Beta, where the scale (δ) and shape parameters (ρ, and a and b) are functions of the global MOE of each board, $E_{t,0,\text{glob}}$. These four models are presented in Table 5.11.

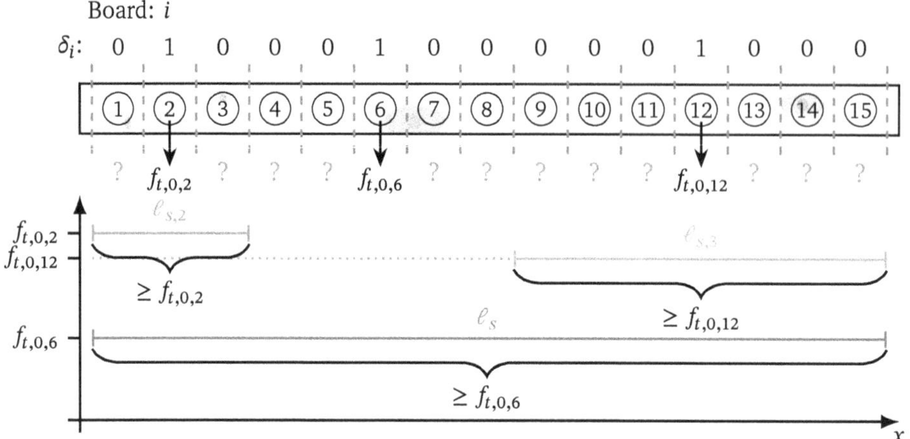

Figure 5.17. Assumed censored data in the tensile tests of boards (compare with Fig. 5.3)

The Weibull distribution is a common choice to describe the strength of different materials—especially when brittle failure is observed—, as it was developed with the concept of *weakest link* (WL) in mind. Including this distribution in the analysis results therefore naturally. The Beta distribution is less commonly used for the description of strength of materials. However, having seen the results of the localized MOE variation and the rather left-skewness of its distribution (see Section 5.4.2), it seems reasonable to test a distribution that is able to express a similar behavior, while at the same time being compatible with the WL theory.

The data fed to the models were constructed in the following way:

1. A binary vector $\underline{\delta}$ of size $15 \cdot N$ (N = total number of tested boards; multiplier 15 represents the number of 100 mm long cells per board) was assembled with either "1" or "0", corresponding to whether a failure has been observed in the cell or not, respectively.

2. A second vector of tensile strengths, $\underline{\boldsymbol{f}}_{t,0}$, of size $15 \cdot N$ is created in the following manner:

   - if $\delta_i = 1$, then the observed $f_{t,0}$ value is inserted,
   - if $\delta_i = 0$, then the value assigned corresponds to the highest tensile strength registered during a test where the i-th cell was part of the free length of the board. This is interpreted as "this cell has *at least* a tensile strength $f_{t,0,i}$".

3. For the case of the models (c) and (d), representing the censored regressions, a third vector was created with the corresponding $E_{t,0,\text{glob}}$ value for each cell, as MOE is assumed constant for each cell in a given board.

Table 5.11. Statistical distributions and fitted parameters for the different investigated models to describe the tensile strength variation within board

Model	Parameters[(1)]
Censored parametric models:	
(a) Weibull	(λ, δ, ρ)
(b) Beta	(λ, δ, a, b)
Censored regressions:	
(c) Weibull	$(\alpha_0, \alpha_1, \beta_0, \beta_1)$
(d) Beta	$(\alpha_0, \alpha_1, \beta_0, \beta_1, \gamma_0, \gamma_1)$
with:	$\lambda = 0$ $\delta = \exp\left(\alpha_0 + \alpha_1 \cdot E_{t,0,\text{glob}}\right)$ $\rho, a = \exp\left(\beta_0 + \beta_1 \cdot E_{t,0,\text{glob}}\right)$ $b = \exp\left(\gamma_0 + \gamma_1 \cdot E_{t,0,\text{glob}}\right)$
Distributions:	
Weibull:	$F(x) = 1 - \exp\left[-\left(\frac{x-\lambda}{\delta}\right)^{\rho}\right]$
Beta:[(2)]	$f(x) = \frac{\Gamma(a+b)}{\Gamma(a)\Gamma(b)} \cdot \left(\frac{x-\lambda}{\delta}\right)^{a-1}\left(1 - \frac{x-\lambda}{\delta}\right)^{b-1}$

[1:] λ: location; δ: scale; (ρ, a, b): shape parameters.
[2:] The probability density function (PDF) is presented instead of the cumulative density function (CDF), since the Beta distribution has no closed expression for its CDF.
[*] $\Gamma(\cdot)$: Gamma function

For the application of the above described method, the python library *Lifelines* (Davidson-Pilon et al., 2019) was used, which implements the needed framework for the analysis of censored data. The models were implemented according to the specifications of *Lifelines*, and are available in Tapia and Aicher (2021).

Survival analysis has been used before for a similar study by Fink (2014), however the methodology used there was very different. Specifically, the likelihood function was not considered and instead an iterative method was applied to fit a parametric Normal distribution, according to Chatterjee and McLeish (1986).

Parametric model

Table 5.12 presents the estimated parameters of the two studied parametric strength distribution models, which were fitted to the tensile strength data, $f_{t,0,\text{cell}}$, of the whole set ($N = 47$) and to the subset of grades LS10 and LS13 boards ($N = 39$). Further, the $\log \mathcal{L}$ and the Akaike information criterion (AIC) (Akaike,

1974) are given for each model. Figure 5.18 shows the two fitted distributions for the whole set.

Considering all boards, the two parametric models, (a) Weibull and (b) Beta, result in very similar $\log \mathcal{L}$ values, meaning that the quality of the fitting process is comparable. However, the Beta model is penalized by the AIC owed to the one extra parameter ($\mathrm{AIC} = 2k - 2\log \mathcal{L}$, with k number of parameters). Figure 5.18a shows both fitted parametric models as compared to the non-parametric Kaplan-Meier (KM) estimator (Kaplan and Meier, 1958). A good agreement can be observed between both model curves and the KM estimator. However, beyond the lower tail of the distributions both models behave fundamentally different, which is apparent from the respective PDFs in Fig. 5.18b. The most important difference in this regard is that the Beta model imposes an upper limit, estimated for the given case as loc + scale $\approx$ 100 MPa. In contrast, the Weibull model is not capped, meaning that much higher local tensile strengths should be possible. It is also clear that the fitted Beta distribution presents a rather marked skewness "to the left", associated with an abrupt descent of the right tail of the PDF. Contrary, the estimated Weibull model shows a more symmetric behavior, even slightly right-skewed, reaching higher $f_{t,0,\mathrm{cell}}$ values.

Similar observations can be made for the behavior of the fitted parameters for the LS10+LS13 subset, revealed graphically in Fig. 5.19. Here, however, the Beta distribution presents a lower AIC value than the Weibull model. While the Weibull model (a) behaves very similarly as in the case of all boards, the Beta model (b) exhibits a sharp end, determined by the shape parameter $b = 1$.

The main difference between the Beta and the Weibull distributions is that the Beta distribution is bounded from both sides, while the Weibull distributions are unbounded to the right. This has a noticeable effect in the shape of the distribution when maximizing the $\log \hat{\mathcal{L}}$ in Eq. (2.9). Since it is reasonable to imagine a finite maximum possible tensile strength for the boards, the double-sided bounding can

Table 5.12. Fitted parameters and statistical information for the different fitted distributions

		loc	scale	shape parameters		$\log \hat{\mathcal{L}}$	AIC
	Model	λ	δ	ρ, a	b		
All boards:							
(a)	Weibull	5.92	74.5	2.39	–	−600.6	1207.3
(b)	Beta	6.93	93.4	2.15	1.23	−600.3	1208.5
LS10+LS13 grades:							
(a)	Weibull	8.71	70.4	2.31	–	−491.6	989.1
(b)	Beta	10.6	79.3	1.97	1.00	−490.1	988.1

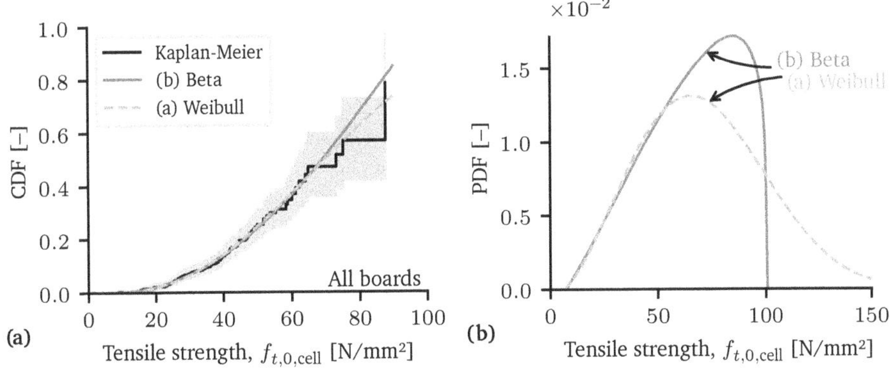

Figure 5.18. Fitted distributions for $f_{t,0,\text{cell}}$ of all boards (a) CDFs and Kaplan-Meier estimator; (b) PDFs.

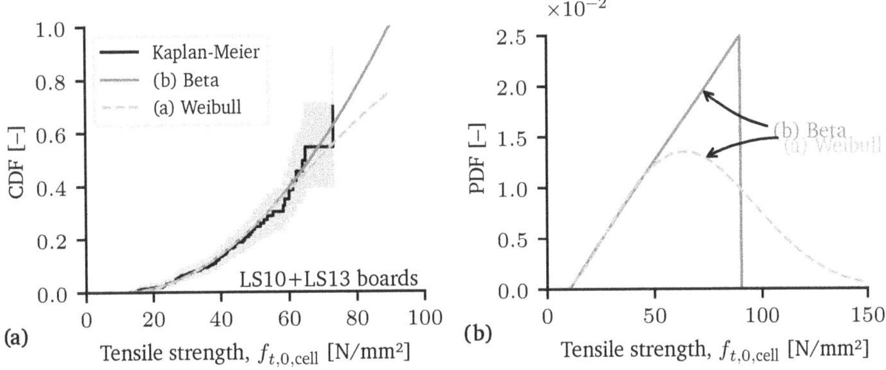

Figure 5.19. Fitted distributions for $f_{t,0,\text{cell}}$ for LS10+LS13 boards(a) CDFs and Kaplan-Meier estimator; (b) PDFs;

be regarded as a reasonable property in this case—preventing negative values on the one side and unrealistically large values on the other. The maximum value for the Beta distribution ("loc" + "scale" = 100.3 N/mm^2) resulted directly from the optimization process (no parameters were fixed). However, if there was a reason to assume a specific maximum value (e.g. due to results of a different batch of specimens), then this could be considered in the fitting process by setting the "scale" parameter accordingly.

Although this analysis provides a good insight into the $f_{t,0,\text{cell}}$ variation, a further step can be made in order to differentiate the level of variation according to a global indicator of each board. The results of such approach are presented next.

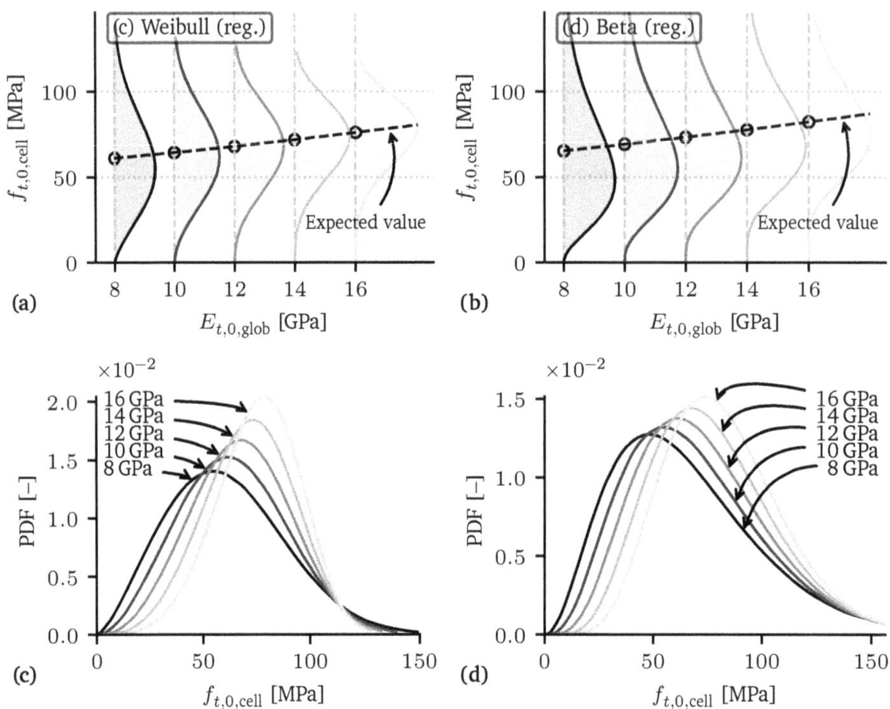

Figure 5.20. Parametric $f_{t,0,\mathrm{cell}}$–$E_{t,0,\mathrm{glob}}$ regression models fitted to all boards. Probability distribution functions drawn for percentiles between 0 % and 99.8 % (a,b) strength–MOE relationship and PDFs for different MOEs; (c,d) PDFs of $f_{t,0,\mathrm{cell}}$ distributions for constant MOE values

Parametric regression model

The censored regression analysis was performed with both models (c) Weibull and (d) Beta, considering the global MOE as the explanatory variable. Similarly as for the parametric models, the analysis was performed for all boards and for the subgroup of LS10+LS13 boards. Due to numerical problems the Beta regression model could not be fitted to the data subset of LS10 and LS13 boards.

Figures 5.20a and 5.20b illustrate the results of the fitted Weibull and Beta regression models for all grades, by means of several PDF curves for ascending $E_{t,0,\mathrm{glob}}$ values in the range of 8 GPa to 16 GPa. Figures 5.20a and b show the PDFs of the local strength distributions for selected, constant MOE values. Figures 5.21a and b analogously reveal the results for the strength grade sub-group LS10+LS13. The regression lines in Figs. 5.20a and 5.21a reveal a clear increase in the expected value of $f_{t,0,\mathrm{cell}}$ is observed from both models with growing $E_{t,0,\mathrm{glob}}$. Thus, these two models capture both, the variation of $f_{t,0,\mathrm{cell}}$ and its correlation with the global MOE of boards.

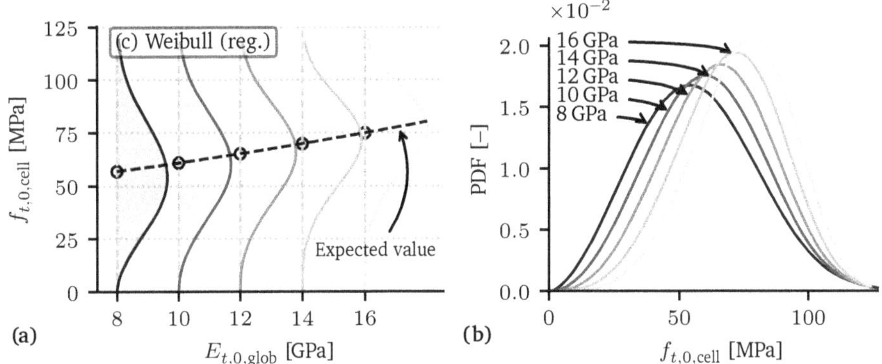

Figure 5.21. Parametric $f_{t,0,\text{cell}}$-$E_{t,0,\text{glob}}$ regression models fitted to the subgroup LS10+LS13. Probability distribution functions drawn for percentiles between 0 % and 99.8 % (a) strength–MOE relationship and PDFs for different MOEs; (b) PDFs of $f_{t,0,\text{cell}}$ distributions for constant MOE values

Table 5.13 presents the estimated parameters for the fitted Weibull and Beta regression models. Model (c) starts with a rather right-skewed and wide-spread distribution for lower $E_{t,0,\text{glob}}$ (skewness $s = 0.42$ for $E_{t,0,\text{glob}} = 8$ GPa, see Fig. 5.20c), and evolves towards a more symmetric, lower variation distribution for higher $E_{t,0,\text{glob}}$ values ($s = -0.18$ for $E_{t,0,\text{glob}} = 16$ GPa). Model (d) starts with a wide-spread, right-skewed distribution for lower $E_{t,0,\text{glob}}$ values, too, however its shape is kept rather constant for higher $E_{t,0,\text{glob}}$, but shifted upwards towards higher $f_{t,0,\text{cell}}$ values (see Fig. 5.20d). Paradoxically, the regression model (c) resembles more the behavior of the parametric Beta model (b), whilst model (d) (Beta) behaves more similarly to the parametric model (a) (Weibull). Table 5.14 shows the log-likelihood and AIC associated to each fitted model, where it can be observed that the AIC is lower for the Weibull model, i.e. the Weibull model should be preferred.

It is worth noting, that models considering the location parameter λ also as a free parameter (i.e. $\lambda \neq 0$, both dependent and independent of $E_{t,0,\text{glob}}$) did not render satisfactory results, but in fact resulted in mechanically unlikely distributions. This could be related to an insufficient number of data points, or to numerical instabilities, preventing to find the optimum result. For the case of highly non-linear problems with many parameters to fit, the choice of the right initial values has an important influence on the solutions. For the analyzed models with $\lambda = 0$ the estimated parameters represent the optimal ones, which was proven by checking multiple different combinations of the initial conditions.

Table 5.13. Estimated parameters for the regression models Weibull and Beta

	Model	loc λ	scale δ	shape parameters ρ, a	b
All boards:					
(c)	Weibull	0	α_0: 4.04	β_0: $2.22 \cdot 10^{-1}$	–
			α_1: $2.36 \cdot 10^{-5}$	β_1: $8.01 \cdot 10^{-5}$	–
(d)	Beta	0	α_0: 5.90	β_0: $2.02 \cdot 10^{-1}$	γ_0: 2.01
			α_1: $5.50 \cdot 10^{-5}$	β_1: $1.19 \cdot 10^{-4}$	γ_1: $1.48 \cdot 10^{-4}$
LS10+LS13 grades:					
(c)	Weibull	0	α_0: 3.91	β_0: $4.84 \cdot 10^{-1}$	–
			α_1: $3.11 \cdot 10^{-5}$	β_1: $6.34 \cdot 10^{-5}$	–

Table 5.14. Log-likelihood and AIC for the fitted parameters of the Weibull and Beta regression models

	Model	$log\hat{\mathcal{L}}$	AIC
All boards:			
(c)	Weibull	−591.8	1192
(d)	Beta	−590.5	1193
LS10-LS13 grades:			
(c)	Weibull	−484.9	978

Verification of parametric models by means of Order Statistics

The goodness of the fitted Weibull (a) and Beta (b) parametric models can be checked by extreme value theory. As previously mentioned, if the fitted models are describing the tensile strength variation along the board, then the application of Eq. (2.6) to the CDF of the fitted models should match the experimental values of $f_{t,0,\mathrm{glob}}$ (minimum value of each board). This is demonstrated in Figs. 5.22b and 5.22a for the Weibull and Beta models, respectively.

It is apparent that the theoretical minimum CDF curve $F_{\min}(x)$ matches rather exactly the experimental results using an exponent $n = 15$ in Eq. (2.6). This is the desired behavior in a stochastic model for $f_{t,0,\mathrm{cell}}$, as this means that the model should be able to capture the size effect (see below) in a rather good manner. Regarding the censored regression models (c) and (d), a similar analysis could not be performed, as their parameters depend on $E_{t,0,\mathrm{glob}}$. This would need a comparison against experimental data for different, specific $E_{t,0,\mathrm{glob}}$ values, which is not possible for this dataset due to the rather small sample size.

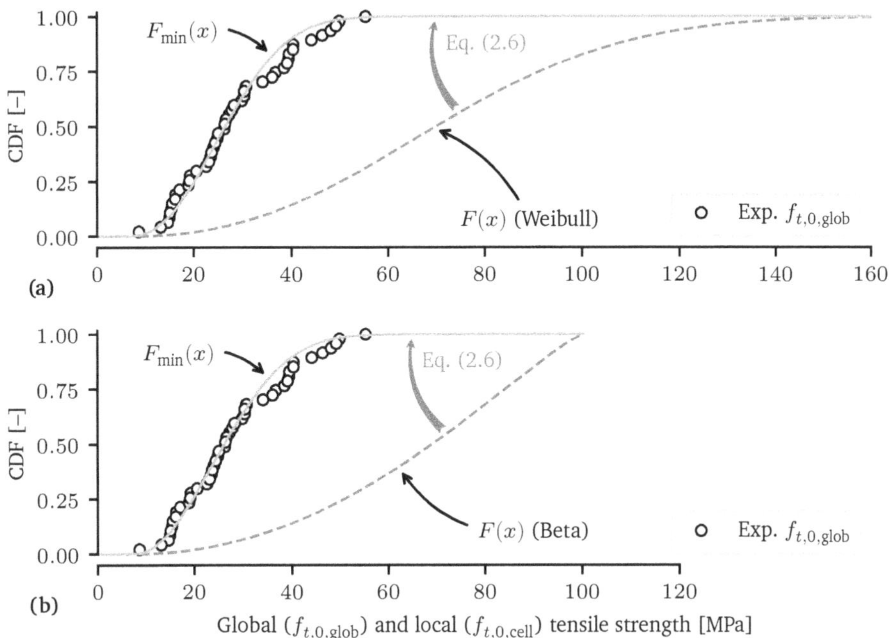

Figure 5.22. Application of Eq. (2.6) to two of the fitted parametric models: (a) Weibull model; (b Beta model). The shown experimental values correspond to the tensile strength per board ($f_{t,0,\mathrm{min}}$)

5.6 Regression models for MOE and $f_{t,0}$

The presented analysis of the variation of mechanical properties within board presented in the thesis is quite different to the more usual approach consisting on the use of linear regressions. Nevertheless, it is deemed important to produce results in line with the proposed models of Ehlbeck and Colling (1987), Blaß et al. (2005), and Fink (2014), too. This allows to assess the suitability of these models for the studied oak material.

Table 5.15 presents a summary of the results obtained by applying different regression models from the literature to the experimental data. These models take the same form as the regression models presented in Table 2.1. The results are briefly discussed in the following.

Firstly the MOE results obtained with Eqs. (a) and (c) of Table 5.15 used by Ehlbeck and Colling (1987) and Blaß et al. (2005) are discussed. It can be observed that, although Eq. (c) adds an additional density term (ρ_0^2), no effect is observed in the computed R-value. However, a noticeable effect is observed for the AIC, which clearly indicates the usefulness of the additional term. Here, the global density

Table 5.15. Results for different literature-proposed regression models for MOE and tensile strength at the global, clear wood and weak section level

Model		Parameters	R	AIC	N
Karlsruher Rechenmodel: (Ehlbeck and Colling, 1987)					
$\ln(E_{t,0}) = \beta_0 + \beta_1 \text{KAR} + \beta_2 \rho_0 + S(0,\sigma)^a$	(a)	$\beta_0 = 9.44$ $\beta_1 = -1.49$ $\beta_2 = -2.39\times10^{-4}$ $\sigma = 2.05\times10^{-1}$	0.52	−228.1	705
$\ln(f_{t,0}) = \beta_0 + \beta_1 \ln(E_{t,0}) + \beta_2 \text{KAR} \cdot \ln(E_{t,0}) + S(0,\sigma)^b$	(b)	$\beta_0 = 3.78\times10^{1}$ $\beta_1 = -3.67$ $\beta_2 = -1.20$ $\sigma = 3.58\times10^{-1}$	0.55	85.6	102
Karlsruher Rechenmodel: (Blaß et al., 2005)					
$\ln(E_{t,0}) = \beta_0 + \beta_1 \text{KAR} + \beta_2 \rho_0 + \beta_3 \rho_0^2 + S(0,\sigma)^a$	(c)	$\beta_0 = 9.45$ $\beta_1 = -1.51$ $\beta_2 = -1.61\times10^{-3}$ $\beta_3 = 1.94\times10^{-6}$ $\sigma = 2.04\times10^{-1}$	0.52	−234.7	705
$\ln(f_{t,0}) = \beta_0 + \beta_1 E_{t,0} + \beta_2 E_{t,0} \cdot \text{KAR} + S(0,\sigma)^b$	(d)	$\beta_0 = 2.54$ $\beta_1 = 1.10\times10^{-4}$ $\beta_2 = -1.38\times10^{-4}$ $\sigma = 3.58\times10^{-1}$	0.55	86.0	102

(Continued on next page)

Table 5.15. *(Continued from previous page)* Results for different literature-proposed regression models for MOE and tensile strength at the global, clear wood and weak section level

Model		Parameters	R	AIC	N
Model by Fink (2014):					
$\ln(E_{t,\mathrm{CWS}}) = \beta_0 + \beta_1 E_{\mathrm{dyn,F}} + S(0,\sigma)$ [c]	(e)	$\beta_0 = 8.56$ $\beta_1 = 7.08\times10^{-5}$ $\sigma = 1.19\times10^{-1}$	0.79	−799.4	565
$\ln(E_{t,\mathrm{WS}}) = \beta_0 + \beta_1 E_{\mathrm{dyn,F}} + \beta_2 \mathrm{KAR} + S(0,\sigma)$ [d]	(f)	$\beta_0 = 8.76$ $\beta_1 = 6.03\times10^{-5}$ $\beta_2 = -1.89$ $\sigma = 2.52\times10^{-1}$	0.61	17.4	140
$\ln(f_{t,\mathrm{board}}) = \beta_0 + \beta_1 E_{\mathrm{dyn,F}} + \beta_2 \mathrm{KAR} + S(0,\sigma)$	(g)	$\beta_0 = 2.72$ $\beta_1 = 6.19\times10^{-5}$ $\beta_2 = -1.80$ $\sigma = 3.14\times10^{-1}$	0.63	31.4	49
$\ln(f_{t,\mathrm{WS}}) = \beta_0 + \beta_1 E_{\mathrm{dyn,F}} + \beta_2 \mathrm{KAR} + S(0,\sigma)$ [b,d]	(h)	$\beta_0 = 3.12$ $\beta_1 = 4.57\times10^{-5}$ $\beta_2 = -2.07$ $\sigma = 3.27\times10^{-1}$	0.55	36.1	50

[a] global density data used
[b] the data of every failed cell was considered
[c] clear wood section (CWS) defined as cells with $\mathrm{KAR} \leq 0.05$
[d] weak sections (WS) defined as cells with $\mathrm{KAR} > 0.05$

was used for the computations, as many cells would remain unconsidered if only local densities would have been used (local densities are not available for every cell). Nevertheless, if only the local densities are used, correlations R=0.57 and R=0.58 are obtained for models (a) and (c), respectively.

Regarding the results for the $f_{t,0}$ models of Eqs. (b) and (d) it can be concluded that there is no noticeable difference in their prediction quality. Both models (Eqs. (b) and (c)) show equal R-values R=0.55, and very similar AIC values of 85.6 and 86.0, respectively.

The application of the models (e)–(h) from Fink (2014) to the oak data produces in general better correlations than the models (a)–(d). For example, a high R-value of 0.79 is obtained for Eq. (e), however, the R-value diminishes to 0.61 for (f) where only weak sections are considered. Since only a subset of the data was used to fit Eqs. (e) and (f) (CWS and WS respectively) a direct comparison with the results from Eqs. (a) and (c) is not possible. However, if all the cells are used in Eq. (f) (N=705) correlation coefficients R=0.76 and AIC=−622.7 are obtained. This means that by all measures that the regressions by Fink (2014) deliver better results for $E_{t,0}$ than the above mentioned models. This is to be expected, as the dynamic modulus of elasticity presents in general a high correlation with the global MOE.

5.7 Summary and discussion

The within-board variation of MOE and tensile strength were discussed in detail, and to a minor extent the effect of density, too. To study the MOE variation, a two-step transformation process was applied to the empirical cell MOE data. In the first step, the empiric MOEs are normalized by the averaged intraboard maxima, and the resulting data are used to fit a statistical distribution. Due to the specific hardwood-growth characteristics in the regarded case (European white oak), the normalized cell data present clear left skewness, which is best approximated by a log-gamma distribution. Finally, the normalized MOE values are then mapped onto a $\mathcal{N}(0,1)$ normal distribution, thus achieving stationarity.

The left skewness of the (normalized) MOE is owed to the localized low MOE values attributed to the detrimental effect of local imperfections, such as knots and fiber deviations. It can be proven that the normalization method itself preserves the original shape of the distribution, which gives a solid base to assume that the obtained left-skewed distributions within board accurately represent the variation of MOE.

Furthermore, the distribution for $\widetilde{E}_{t,0,\text{cell}}$ should, in theory, be determined for each studied batch of boards. An experimental campaign using the here reported methodology can be very time-consuming. Thus, ideally this distribution would be parameterized in terms of certain characteristics of the boards, like grading class or origin. Naturally, this would require a large enough dataset in a first place. Since the presented method is rather slow, different techniques could be used to accelerate the process of gathering the needed data. Possible methods could be digital image correlation or fiber-optic strain measurements.

The application of survival analysis to the study of the variation of $f_{t,0,\text{cell}}$ proved to deliver reasonable results, which are in agreement with the extreme value theory. This means that a model based on this results should be capable of simulating a size effect. This will be investigated in the next chapter. Whether the Weibull or Beta distribution is more suitable for the simulation of $f_{t,0,\text{cell}}$ values is a more difficult question. Considering the AIC's from Table 5.12, the Weibull model should be chosen, as it represents the data with higher probability than the Beta distribution. Nevertheless, the fitted Beta model is appealing, too, since it considers a clear upper limit for the tensile strength within a single board.

The two analyses carried out for MOE and $f_{t,0}$ produced the needed components for the development of a material simulation model. Such model should be capable to reproduce the observed properties in a probabilistic sense, making it suitable to be used as input for stochastic models of glulam.

Simulation of mechanical properties along oak boards

6.1 General remarks

This chapter presents models for the simulation of mechanical properties along boards required for glulam strength models, such as the model introduced in Chapter 7. The models comprise the generation of MOE, tensile and compressive strength profiles along boards, while preserving the statistical characteristics observed at the global level (i.e. according to standardized tests). This is done by applying the results from the previous chapter regarding the autocorrelation of the variation of MOE within board, as well as the intra-board tensile strength distributions.

The general procedure to simulate the mechanical properties can be divided in two main subtasks: (i) The simulation of correlated global properties ($E_{t,0,\mathrm{glob}}$, $f_{t,0,\mathrm{glob}}$ and $f_{c,0,\mathrm{glob}}$), and (ii) the generation of profiles of the corresponding localized mechanical properties ($E_{t,0,\mathrm{cell}}$, $f_{t,0,\mathrm{cell}}$ and $f_{c,0,\mathrm{cell}}$). Figure 6.1 illustrates these steps.

The global properties are generated considering the correlation matrix, while the simulation of localized profiles is based on an autoregressive model. For the case of the localized MOE, the procedure considers the generation of stationary processes, which are then translated to MOE profiles by reversing the two-step

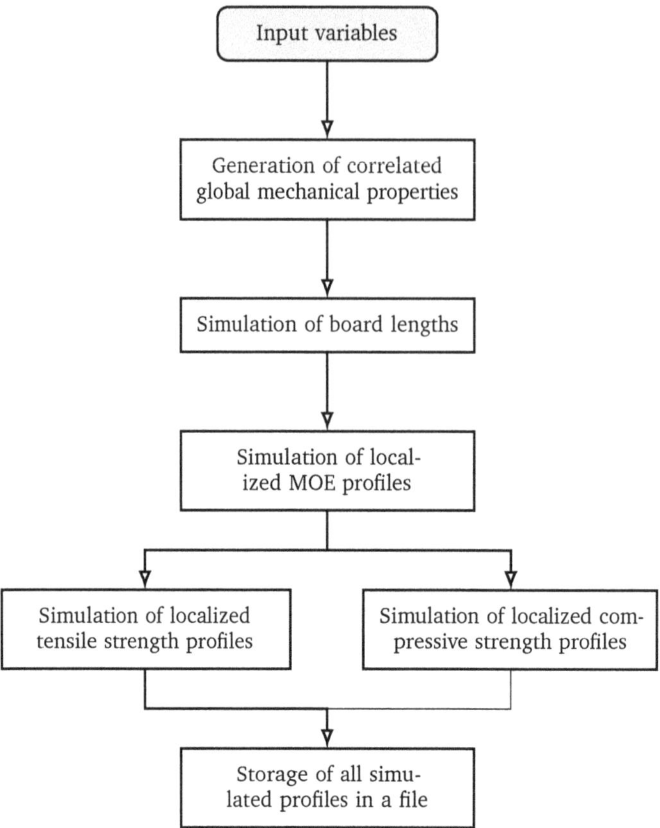

Figure 6.1. General flowchart of the simulation of localized mechanical properties

method described in the previous chapter. Tensile and compressive strength profiles are simulated based on the same stationary data, too. Here, the cross-correlation between the different mechanical properties at a local level is considered by a cross-correlation coefficient. The details of each step are presented in the following.

6.2 Simulation of global properties of boards

For the simulation of global mechanical properties, the method described in 2.4.1 is applied. This approach uses the concept of the multivariate normal distribution to preserve the correlation between variables. Additionally, it considers the mapping of the correlated variables from the $\mathcal{N}(0,1)$ space onto the corresponding distribution for each global variable. The mechanical properties regarded here are the MOE parallel to the main axis of the board, $E_{t,0,\mathrm{glob}}$, the tensile strength parallel to the grain, $f_{t,0}$, and the compressive strength parallel to the grain, $f_{c,0}$. The input data needed by the model is obtained by means of the commonly used,

Table 6.1. Correlation matrix, Σ, for the mechanical strength-MOE properties of dataset B

	$E_{t,0,\text{glob}}$	$f_{t,0}$	$f_{c,0}$
$E_{t,0,\text{glob}}$	1.00	0.48	0.81
$f_{t,0}$	0.48	1.00	0.80
$f_{c,0}$	0.81	0.80	1.00

Table 6.2. Parameters for the global mechanical properties of the boards

Variable	Distribution	loc	scale	shape
$E_{t,0,\text{glob}}$	lognorm	6.22×10^3	6.58×10^3	4.20×10^{-1}
$f_{t,0}$	lognorm	-3.52×10^2	8.88×10^2	1.83×10^{-1}
$f_{c,0}$	lognorm	-2.26×10^{-1}	5.03×10^2	4.62×10^{-2}

standardized tests, for example according to EN 408 (2012).

The application of this method is demonstrated taking as input parameters the statistical information of Dataset B (specifically the combined grade LS10+LS13, see Table 3.3). For the oak boards belonging to this dataset the correlation matrix Σ is presented in Table 6.1. The correlation between $f_{t,0}$ and $f_{c,0}$ has not been determined empirically, and was assumed as R=0.8. However, it is noteworthy to mention that this value is only mathematically needed and does not influences the relevant generated variables significantly: the correlation $f_{t,0}$–$f_{c,0}$ has no relevance, as both variables are mutually exclusive in a GLT model: either $f_{t,0}$ or $f_{c,0}$ is assigned depending on the position of the board (see Fig 7.2). The experimentally estimated parameters for the statistical distributions of these variables are shown in Table 6.2.

Figure 6.2a shows the simulated global values for $E_{t,0,\text{glob}}$ and $f_{t,0}$, for a total of 5000 boards. The simulated data fits the input distribution very closely. At the same time, the correlation R=0.48 between both variables is preserved, too. The same conclusion can be drawn from Fig. 6.2b, where the simulated $E_{t,0,\text{glob}}$ and $f_{c,0}$ are compared. In this case, the correlation obtained is slightly smaller as specified in the covariance matrix, Σ, (0.79 vs. 0.81). This difference in the correlation is owed to the translation $y = F^{-1}(\Phi(x))$. For the studied distributions, this difference can be considered to be very small.

6.3 Localized modulus of elasticity

The model for the simulation of $E_{t,0,\text{cell}}$ values along boards is based on an autoregressive (AR) model that preserves the serial correlation determined in Section 5.4.6. In essence, the steps described in 5.4.6 to analyze the MOE variation

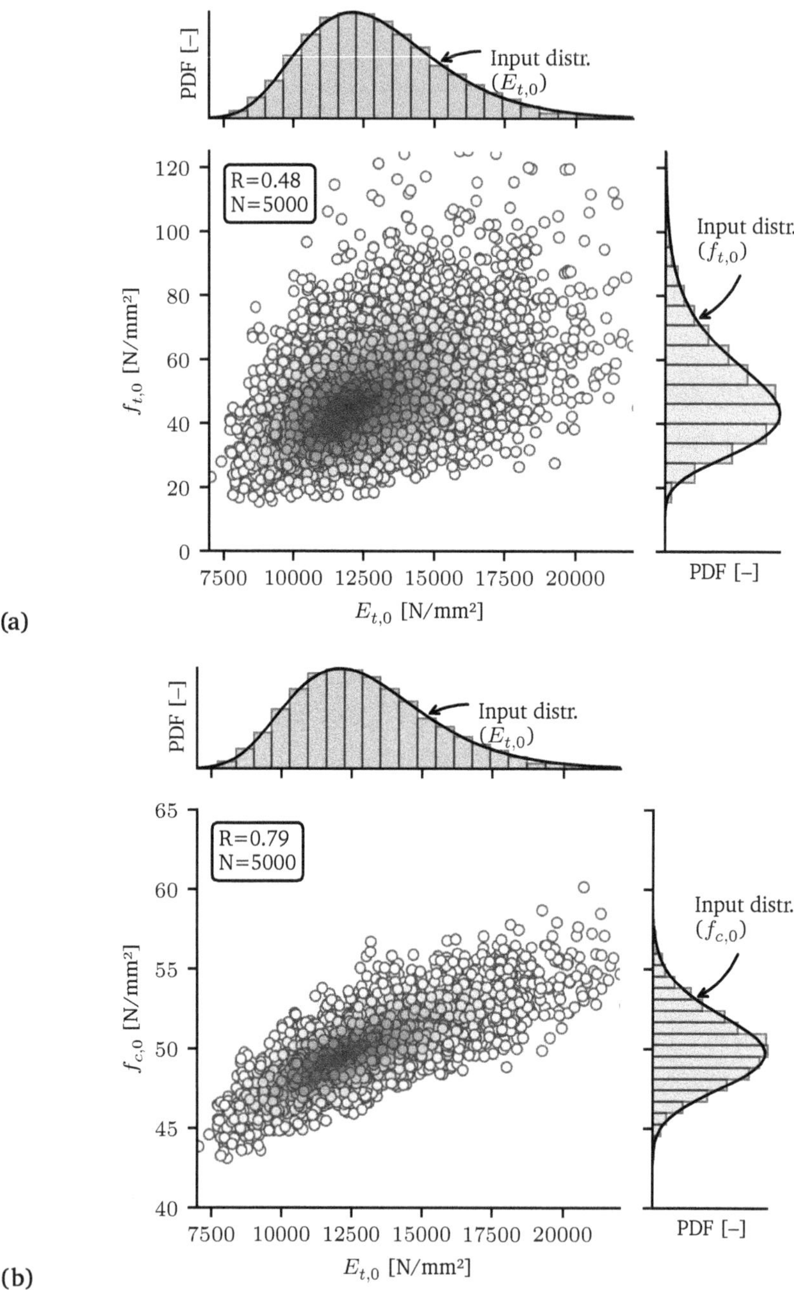

Figure 6.2. Simulated data for global MOE vs. strength values (a) $E_{t,0,\text{glob}}$ vs. $f_{t,0,\text{glob}}$, and (b) $E_{t,0,\text{glob}}$ vs. $f_{c,0,\text{glob}}$

are followed in reverse order. First, (i) a stationary process is generated by means of an AR model, (ii) then this process is mapped into the distribution corresponding to $\widetilde{E}_{t,0,\text{cell}}$, and finally (iii) a scaling factor is applied to these values to *de-normalize* the $\widetilde{E}_{t,0,\text{cell}}$ values and obtain *real* local MOEs for any given global MOE for a board.

In the following, the characteristics of the autoregressive process are analyzed and later the detailed steps to simulate the $E_{t,0,\text{cell}}$ profiles are presented.

6.3.1 Determination of order and parameters of the stationary AR process

The first step for the simulation of MOE profiles along board concerns the sampling of stationary $Z_{t,0,\text{cell}}$ values. The analysis performed in Section 5.4.6 provides all required information for this. The results obtained for the first five lag-correlations indicate, that a first order AR model describes the variation of $Z_{t,0,\text{cell}}$ in a satisfactory manner. This applies for both studied cases: for clear wood segments only (KAR < 0.05), and for all segments in each board. A verification of this statement was tested by fitting the $Z_{t,0,\text{cell}}$ data to both, an AR(1) and an AR(2) model of the form

$$Z_{t,0,i} = \sum_{j=1}^{p} \varphi_j \cdot Z_{t,0,i-j} + \varepsilon_i, \tag{6.1}$$

where φ_i are the model parameters, and ε_i, the white noise component, computed according to Eq. (2.30). The model parameters φ_i were determined by solving Eq. (6.1) for $p = 1$ and $p = 2$ (first and second order, respectively). The equation system was solved by means of the ordinary least squares method.

The fitted parameters for the two models [AR(1) and AR(2)], as well as for the two studied cases, considering all cells and only the CW segments, are presented in Table 6.3. It is apparent that the additional parameter in AR(2) has almost no relevance in the prediction quality of the stationary data (see the φ_2 parameters and the R values in Table 6.3). This irrelevance becomes even more clear when the AIC is considered, where in both cases, the AR(1) presents the lowest value. Furthermore, the fact that the difference of the AIC values for AR(1) and AR(2) is exactly 2 means that the extra parameter has no notable effect in the prediction ($\text{AIC} = 2k - 2\log\mathcal{L}$, where $\mathcal{L}$ is the likelihood and k is number of fitted parameters). It should be noted that the same number of data points was used to compute the AIC for AR(1), as was available for the AR(2) fitting process, since the AIC depends on the number of data points used (owing to the likelihood).

Table 6.3. Parameters for autoregressive models of first [AR(1)] and second [AR(2)] order for local MOE for (i) all cells and (ii) for clear wood (CW) cells with KAR < 0.05

Model	φ_1	φ_2	R	Resid	AIC
$AR(1)_{All}$	0.440	–	0.43	0.96	1722
$AR(2)_{All}$	0.446	−0.014	0.43	0.96	1724
$AR(1)_{CW}$	0.584	–	0.60	0.77	637
$AR(2)_{CW}$	0.586	−0.002	0.60	0.77	639

In conclusion, the AR(1) was chosen to describe the stationary process needed for the simulation of localized MOE profiles along board.

6.3.2 Model for the simulation of MOE profiles

The simulation method is based on an AR(1) model and consists of three main steps, which are analogous to the steps described to obtain the stationary data of the localized empiric MOEs, but in a reversed order. The procedure consists of the following steps:

1. Generate an autoregressive process with the model parameters φ_i estimated from solving the Eq. (6.1). Following the above results, an AR(1) model is used to simulate $Z_{t,0,\text{cell}}$ values as

$$Z_{t,0,i} = \varphi_1 Z_{t,0,i-1} + \varepsilon_i \quad , \tag{6.2}$$

   where ε_i is the white noise component $\mathcal{N}(0,\sigma)$, with $\sigma = \sqrt{1-\varphi_1^2}$. To avoid a strong influence from the initial conditions, the first 30 generated values per board are discarded.

2. The generated $Z_{t,0,\text{cell}}$ values are mapped to the distribution $F(x)$ of the normalized cell MOE data, $\widetilde{E}_{t,0,\text{cell}}$ as

$$\widetilde{E}_{t,0,\text{cell}} = F^{-1}\left[\Phi\left(Z_{t,0,\text{cell}}\right)\right] \quad , \tag{6.3}$$

   where F^{-1} is the inverse CDF of the distribution describing the normalized MOE data and Φ is the standard normal distribution, $\mathcal{N}(0,1)$, which is the exact inverse of Eq. (5.5).

3. Finally, given a value for the global MOE of the board ($E_{t,0,\text{glob}}$), a factor m_0 is computed to scale $\widetilde{E}_{t,0,\text{cell}}$ as

$$E_{t,0,\text{cell}} = m_0 \cdot \widetilde{E}_{t,0,\text{cell}} \quad , \tag{6.4}$$

where $E_{t,0,\text{cell}}$ is the simulated MOE profile of the virtual board. The factor m_0 is computed as

$$m_0 = \frac{E_{t,0,\text{glob}}}{\widetilde{E}_{t,0,\text{glob,cell}}} \quad , \tag{6.5}$$

where $\widetilde{E}_{t,0,\text{glob,cell}}$ is the result after applying the equation for serially arranged springs to the generated $\widetilde{E}_{t,0,\text{cell}}$ values (Eq. (5.2)). The choice of $E_{t,0,\text{glob}}$ can be randomly sampled from a suitable distribution.

An overview of this process is illustrated in Fig. 6.3 (left branch) as a flowchart. Figure 6.3 also describes the simulation of tensile strength profiles, which is described later in this chapter and depends on the simulation of MOE profiles.

6.3.3 Example of simulations of MOE profiles along boards

For the simulation of MOE profiles of oak boards, the procedure described above in conjunction with Eqs. (6.2) to (6.5) is followed. For the simulation of the $Z_{t,0,\text{cell}}$ values, the parameters φ_1, corresponding to the AR(1) models, are taken from Table 6.3. For the distribution of the normalized MOE, $\widetilde{E}_{t,0,\text{cell}}$, the log-gamma distribution is chosen with the corresponding parameters from Table 5.7. Simulations for both studied cases are performed: (1) with the parameters determined on the basis of all cells and (2) considering only clear wood segments. To make both simulated cases comparable at the board level, the same white noise sequences, ε_i, were used for each case.

Examples of four generated MOE profiles are presented in Figs. 6.4a–d. Focusing first on the solid lines (parameter set 1, all cells), it can be seen that a wide range of *typical* (according to the MOE measurements) MOE profiles can be reproduced.

In detail, Fig. 6.4a shows rather low variation in the MOE, while the profile depicted in Fig. 6.4b presents higher variation, where a steady increase in MOE is observed for approximately 800 mm. One could imagine that such a behavior arises in a real board due to a steady decrease in the fiber inclination, which may

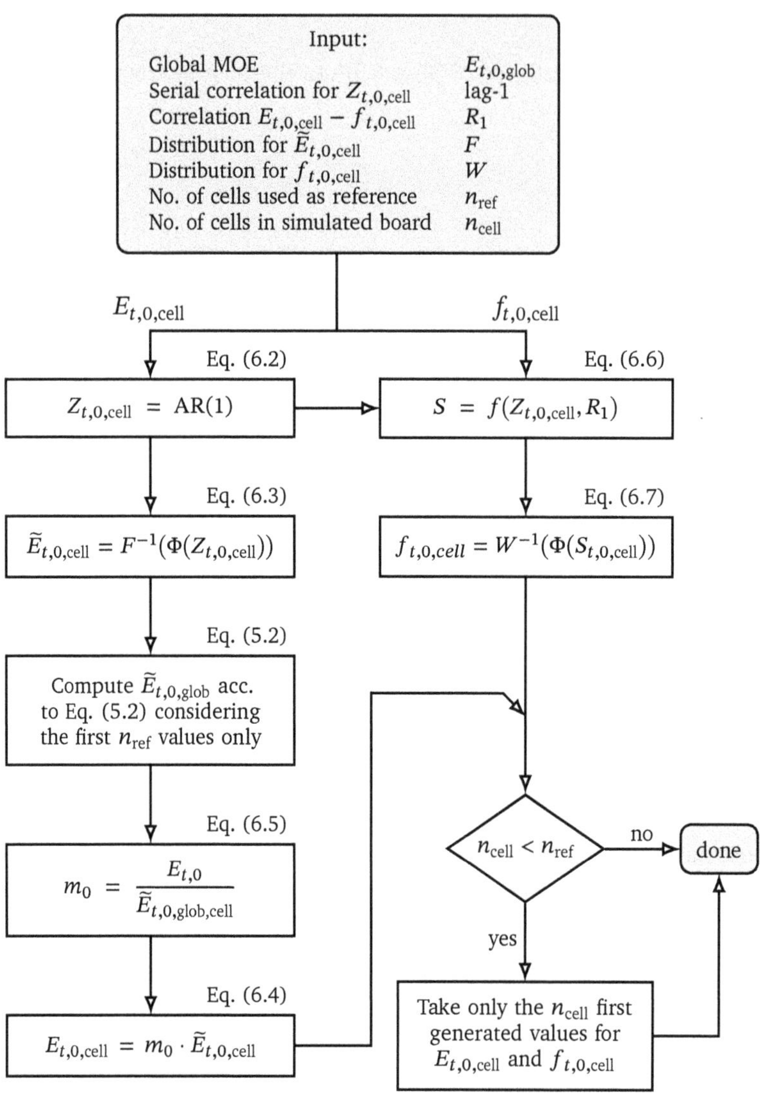

Figure 6.3. Flowchart of the simulation process for localized $E_{t,0,\text{cell}}$ and $f_{t,0,\text{cell}}$ values for a single board

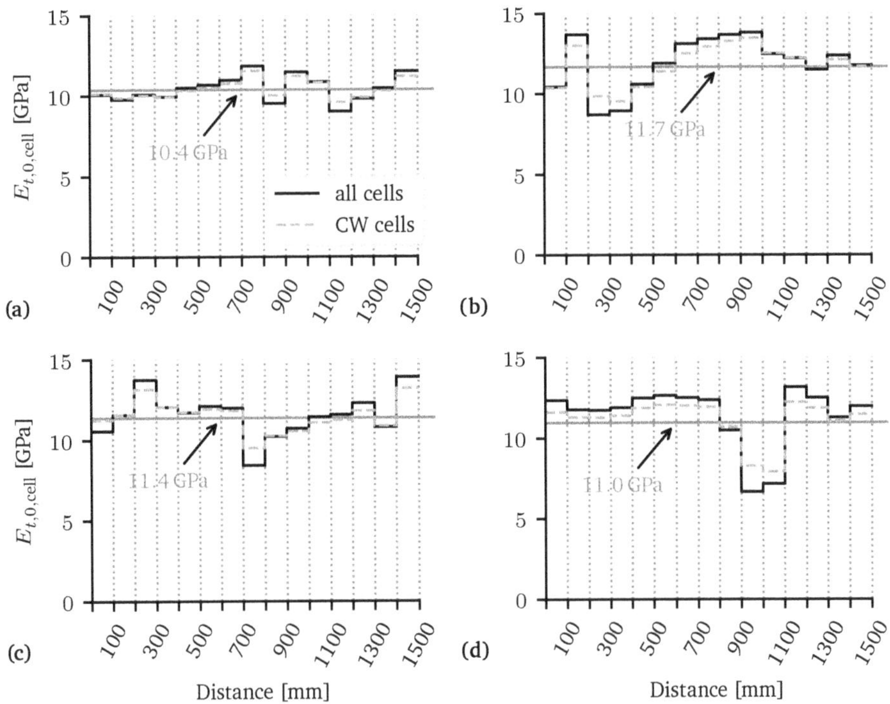

Figure 6.4. Samples of simulated MOE profiles with typical characteristics. (a) low MOE variation, (b) higher localized MOE variation, (c) higher, rather distributed MOE variation, (d) large local defect.

also be related to the presence of knots. Figure 6.4c presents a board with a few downward steps in the MOE in an otherwise rather moderately varying sequence of $E_{t,0,\mathrm{cell}}$ values. This finding would simulate the presence of local defects (e.g., knots). Finally, Fig. 6.4d illustrates the case where a very large local defect is present, considerably lowering the local MOE in a narrow region.

The dashed lines in Figs. 6.4a-d represent the results obtained with parameter set 2 (CW segments). Since the same white noise sequence was used, the curves move very similarly. However, due to the different parameters, some differences in the variation can be observed (see, e.g., Fig. 6.4d). Overall, the variation in the simulated CW segments is slightly smaller than that in all the cells, which can be explained mainly by the difference in the distributions for the normalized MOEs, $\widetilde{E}_{t,0,\mathrm{cell}}$. However, it should be mentioned that the experimental results for the clear wood segments exhibit large variations, which is captured by the fitted $\widetilde{E}_{t,0,\mathrm{cell}}$ distribution. This phenomenon is why clear wood segments do not represent an especially low variation in the studied case—as the concept of "clear wood" might otherwise suggest. In this context, it should be recalled that clear

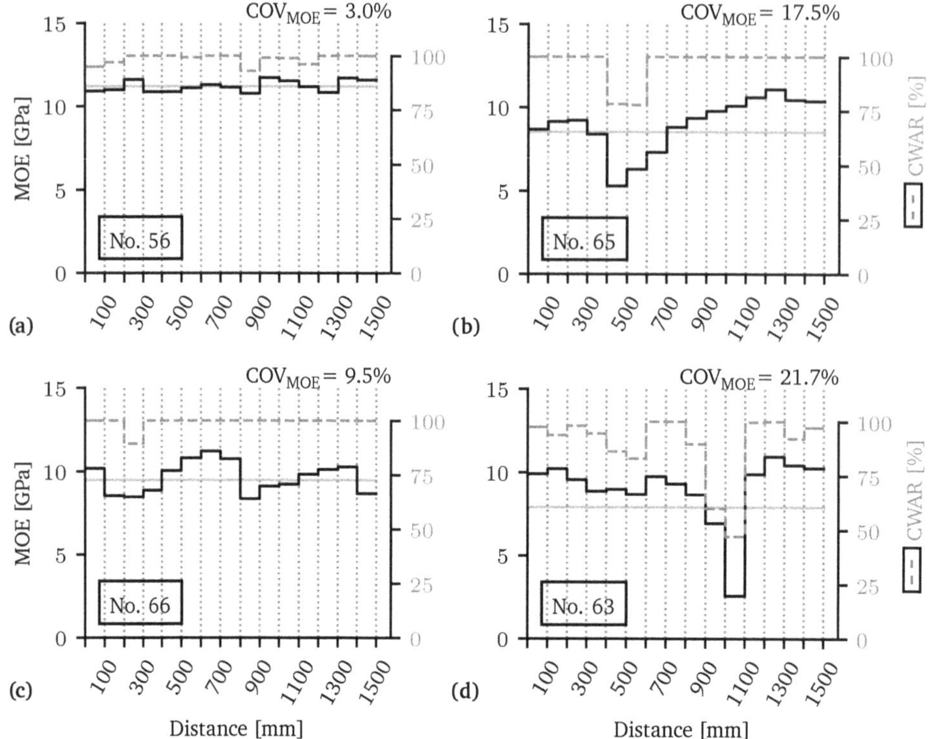

Figure 6.5. Experimentally determined MOE and KAR profiles showing typical characteristics. (a) relatively low variation of MOE, (b) higher MOE variation, (c) higher MOE variation, (d) large local defect.

wood, as defined in this paper, relates exclusively to an almost complete exclusion of knots (KAR ≤ 0.05), but includes any local or global fiber deviation, as it can be assumed that this is the dominant reason for CW MOE fluctuations.

These profiles can be compared to a selection of the empirically determined $E_{t,0,\text{cell}}$ values, presented in Figs. 6.5a–d. Each of these profiles was chosen based on their apparent (visual) similarity with the simulated profiles of Figs. 6.4a-d. In this sense, Fig. 6.5a somehow resembles Fig. 6.4a; Fig. 6.5b is similar to Fig. 6.4b; as well as Figs. 6.5c,d and Figs. 6.4c,d, respectively.

The measured CWAR values (CWAR = 1 − KAR) for each cell are shown in Figs. 6.5a-d, too. It can be seen that in general terms, the simulated profiles present very similar characteristics as the empirical profiles. It should be emphasized that the presented generated profiles are not intended to match the empirical measurements exactly, but rather that the model is capable to reproduce profiles belonging to the same category of variation characteristics. This is achieved by maintaining the autocorrelation function (see Fig. 6.6) and reproducing the distributions of normalized MOE, $\widetilde{E}_{t,0,\text{cell}}$, and global MOE, $E_{t,0,\text{glob}}$.

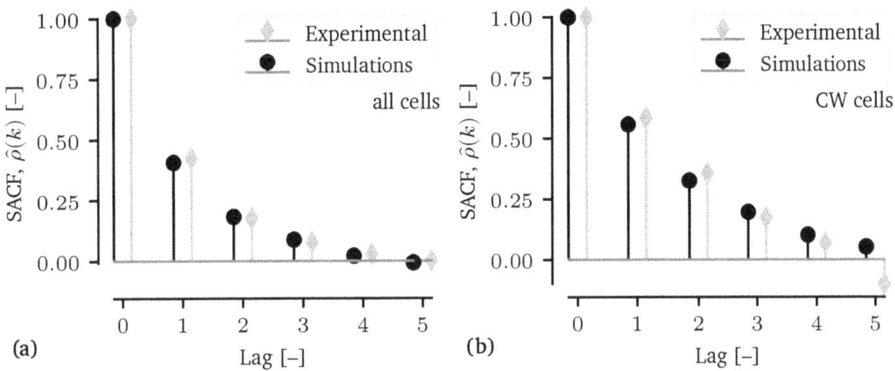

Figure 6.6. Comparison between the computed experimental and simulated SACFs. (a) all cells; (b) CW cells

One further comparison was performed to confirm that the simulated data preserve the specified autoregressive parameters. For this comparison, a single, very long board of $n = 2000$ cells was simulated. Then, the two-step normalization process was applied, and the SACF was computed. Here, the $\bar{E}_{\text{max},3}$ criterion was generalized to use the m highest values of the profile, i.e. $m = \lfloor n \cdot 3/15 \rfloor$, to compensate for the different number of considered cells. The results of this analysis can be seen in Figs. 6.6a and b, for all cells and CW cells, respectively. In both cases, a perfect agreement between the experimental results and the generated profiles can be observed, meaning that the autocorrelation properties are correctly simulated.

6.3.4 Consideration of different board lengths

Since all the measurements were done for a total of 15 cells per board, it follows that the distributions of $\widetilde{E}_{t,0,\text{cell}}$ should only be used for the simulation of boards with the same number of cells—as done in the examples above. However, it is possible to use this model to generate boards with an arbitrary number of cells, where two principal cases can be distinguished. The first one corresponds to the case where a board with a number of cells $n_{\text{cells}} < 15$ needs to be generated. Here, $E_{t,0,\text{cell}}$ values are generated for a total of 15 cells, but only the first n_{cells} values are returned. The second possibility consists on the case where $n_{\text{cells}} > 15$; here, a total of n_{cells} values for $\widetilde{E}_{t,0,\text{cell}}$ are generated, however the scaling according to Eq. (6.4) is applied considering the first 15 cells only. The data simulated in this manner maintains the original statistical characteristics regrading the intra-board variation.

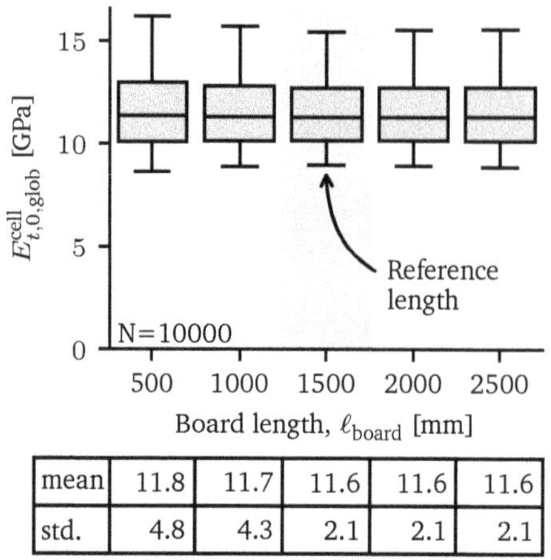

mean	11.8	11.7	11.6	11.6	11.6
std.	4.8	4.3	2.1	2.1	2.1

Figure 6.7. Simulated size effect of $E_{t,0,\text{glob}}$

Figure 6.7 shows the simulation of boards with different lengths (5000 simulations each), where the parameters corresponding to all cells were used. The results present the global MOE computed from the individual cells in each board, $E^{\text{cell}}_{t,0,\text{glob}}$. It can be seen that the mean value remains almost the same for the different simulated lengths, where a minor decrease of 0.2 GPa is observed from 500 mm to 1500 mm. The standard deviation, however, decreases slightly from 3 GPa to 2.1 GPa when moving from 500 mm to 1500 mm. For longer simulated boards (between 1500 mm and 2500 mm) the mean and standard deviation remains practically unchanged.

6.4 Localized tensile strength

The simulation of tensile strength, $f_{t,0,\text{cell}}$, requires the previous generation of MOE profiles along the board, $E_{t,0,\text{cell}}$, as described above. The above presented simulation approach consists on cross-correlating a new stationary process to the generated $Z_{t,0,\text{cell}}$ values and then map the process to the correct strength distribution, $f_{t,0,\text{cell}}$. The direct use of the fitted statistical distribution allows to directly incorporate important statistical features, such as the consideration of the size effect. The latter is shown with simulations in Section 6.4.3. In the following, the model to simulate localized tensile strength profiles is presented in detail and examples are shown.

6.4.1 Model for the simulation of tensile strength profiles

The method to simulate $f_{t,0,\text{cell}}$ values is based in the concept of a vector autoregressive (VAR) model (however simplified), where a set of stochastic processes are cross-correlated. In the given case, a Gaussian process $\mathcal{N}(0,1)$, $S_{t,0,\text{cell}}$, is generated based on a cross-correlation model with the previously simulated $Z_{t,0,\text{cell}}$ values. The values $S_{t,0,\text{cell}}$ are then mapped into the correct distribution for $f_{t,0,\text{cell}}$, as previously estimated with the survival analysis methodology presented in Section 5.5.4. The steps for the simulation of $f_{t,0,\text{cell}}$ are:

1. First, the stationary process $S_{t,0,\text{cell}}$ is generated, which depends on $Z_{t,0,\text{cell}}$ as

$$S_{t,0,i} = \theta_0 \cdot Z_{t,0,i} + \varepsilon_i \quad , \tag{6.6}$$

   where θ_0 is the cross-correlation coefficient between $E_{t,0,\text{cell}}$ and $f_{t,0,\text{cell}}$. In this case the white-noise term has a standard deviation $\sigma_2 = \sqrt{1-\theta_0^2}$. The parameter θ_0 is taken from the results obtained in Section 5.5.3 as $\theta_0 = 0.73$.

2. The vector $\underline{S}_{t,0,\text{cell}}$, which corresponds to all the $S_{t,0,i}$ values in a board, and assumed to belong to a $\mathcal{N}(0,1)$, is mapped into the $f_{t,0,\text{cell}}$ distribution (i.e. any of the fitted distributions $F(x)$ presented in Table 5.11) by means of the transformation

$$\underline{f}_{t,0,\text{cell}} = F^{-1}\left(\Phi\left(\underline{S}_{t,0,\text{cell}}\right)\right) \quad , \tag{6.7}$$

   where F^{-1} is the inverse of the cumulative distribution function (CDF), and Φ is the CDF of the standard Normal distribution. This follows the same principle used for the simulation of MOE profiles, where the data in the standard normal space is transformed into a different distribution.

The vector $\underline{f}_{t,0,\text{cell}}$ obtained in this manner corresponds to the simulated tensile strength profile, which has the following properties: (i) cross-correlation with $E_{t,0,\text{cell}}$ (θ_0 term), (ii) implicit autocorrelation owed to the cross-correlation with $E_{t,0,\text{cell}}$, and (iii) a statistical behavior equal to the one obtained from the survival analysis performed with the experimental data. The described process is presented in Figure 6.3 (right) as a flowchart.

6.4.2 Simulation examples of tensile strength profiles along boards

Simulations of tensile strength profiles were performed with the above described model using the parameters obtained from the studied oak boards. The four fitted models are studied: (a) Weibull, (b) Beta, (c) Weibull regression and (d) Beta regression. To make the results obtained from the different models comparable, the same white noise component ε_i in Eq. (6.6) is used. In addition, the same stationary process $Z_{t,0,\text{cell}}$ is used for each model. Thus, the only difference is the statistical distribution applied in Eq. (6.7).

The simulations are first analyzed for two different simulated boards using the parametric Beta model shown in Figs. 6.8a and 6.8b. Both simulated profiles show a clear correlation between $E_{t,0,\text{cell}}$ and $f_{t,0,\text{cell}}$. Both, the $f_{t,0,\text{cell}}$ and the $E_{t,0,\text{cell}}$ curves move similarly, but differences arise due to the introduced white noise component, ε_i, in Eq. (6.6). In general, regions of lower $E_{t,0,\text{cell}}$ values also present relatively low $f_{t,0,\text{cell}}$ values and vice versa. Note that the $E_{t,0,\text{glob}}$ value, needed for the simulation of the MOE profiles, was taken randomly from the statistical distribution for global MOE from the dataset B (see Section 3.3).

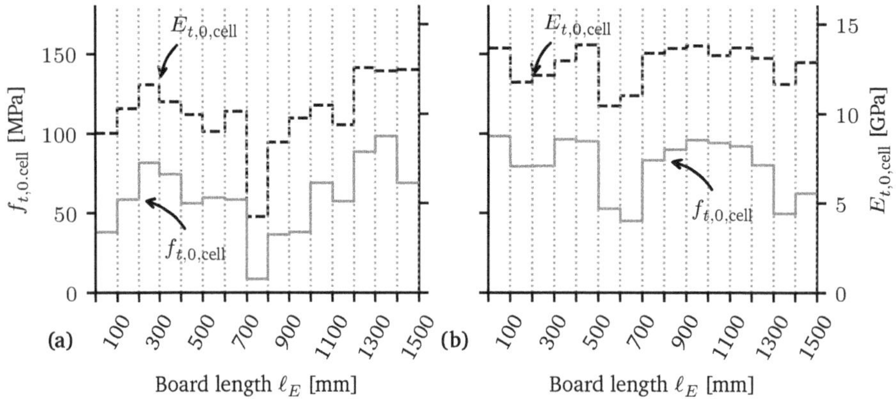

Figure 6.8. Simulation of $f_{t,0,\text{cell}}$ profile with the Beta model (also given are the simulated $E_{t,0,\text{cell}}$ values) (a), (b) boards A and B

Figures 6.9a and 6.9b present the simulation of $f_{t,0,\text{cell}}$ and $E_{t,0,\text{cell}}$ of the same boards A and B as above, however showing all four models. The results of each model are slightly shifted along the horizontal axis, helping to identify each individual model and allowing for a direct comparison between the four models. The blue lines correspond to the parametric models (a) and (b), while the regression models (c) and (d) are drawn in red tones.

The "left" board A (Fig. 6.9a) corresponds to $E_{t,0,\text{glob}} = 9.4\,\text{GPa}$, while the

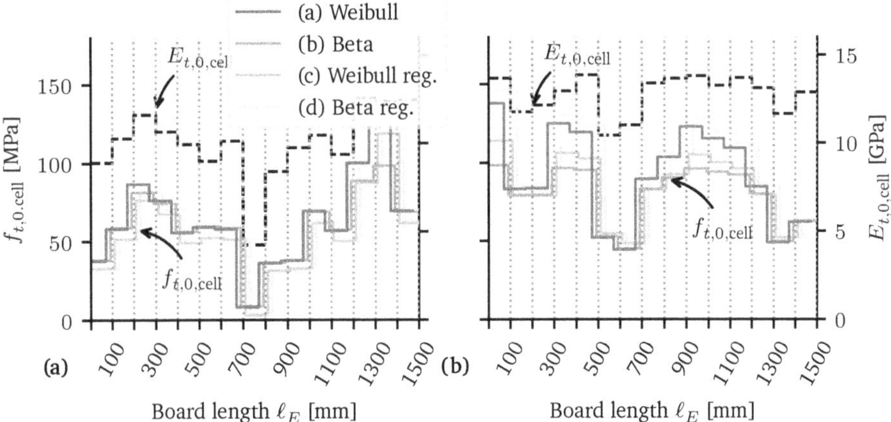

Figure 6.9. Simulation of $f_{t,0,\text{cell}}$ values together with the previously generated $E_{t,0,\text{cell}}$ values; (a), (b) boards A and B

"right" board B (Fig. 6.9b) has a higher global MOE of $E_{t,0,\text{glob}} = 12.6\,\text{GPa}$. It can be seen that for the case of board A (lower $E_{t,0,\text{glob}}$) both regression models show smaller values at the lower end of the profiles. For board B (Fig. 6.9b) the opposite is true. This is directly related to the dependence of $f_{t,0,\text{cell}}$ with $E_{t,0,\text{glob}}$ for the case of the regression models, as boards with higher $E_{t,0,\text{glob}}$ values shift the $f_{t,0,\text{cell}}$ distribution to higher values.

The upper end of the $f_{t,,0,\text{cell}}$ profiles also exhibit some clear differences, where models (a) and (d) clearly present higher values as models (b) and (c). This is, of course, directly related to the characteristics of each statistical model (see Figs. 5.18b and 5.20). While the beta model (b) has a clear maximum possible value at around 100 MPa, the Weibull model (a) allows for higher values of $f_{t,0,\text{cell}}$. A similar relation exists between model (c) and (d), respectively. These differences in the fitted models are expressed directly in the simulated boards.

It is evident that the main disadvantage of the parametric models (a) and (b) is the lack of correlation with the global MOE. Section 6.4.3 presents a simple procedure to consider the correlation with the parametric models, making these a viable alternative to the regression models, which are more complicated to calibrate.

6.4.3 Simulation of length effect for tensile strength

The presented model is capable of simulating the size effect of the tensile strength. This capability derives directly from the "cell"-centered stochastic nature of the model, and is tightly related to the behavior shown in Section 5.5.4 (compliance

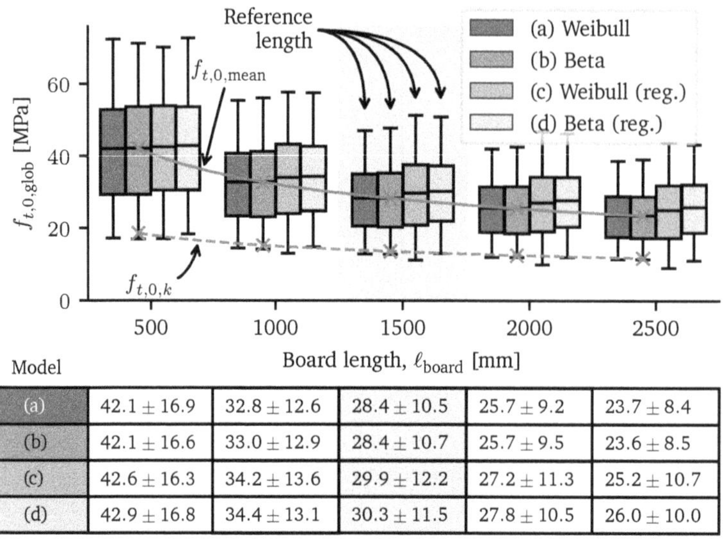

Model	500	1000	1500	2000	2500
(a)	42.1 ± 16.9	32.8 ± 12.6	28.4 ± 10.5	25.7 ± 9.2	23.7 ± 8.4
(b)	42.1 ± 16.6	33.0 ± 12.9	28.4 ± 10.7	25.7 ± 9.5	23.6 ± 8.5
(c)	42.6 ± 16.3	34.2 ± 13.6	29.9 ± 12.2	27.2 ± 11.3	25.2 ± 10.7
(d)	42.9 ± 16.8	34.4 ± 13.1	30.3 ± 11.5	27.8 ± 10.5	26.0 ± 10.0

Figure 6.10. Size effect obtained with the four fitted models for the variation of tensile strength along boards. Mean and standard deviation values for each model and board length are given in tabular form. Fitted size effect law is shown for model (b) for the mean and 5 %-quantile.

with the extreme value theory).

To show the size effect produced by the simulation method, virtual boards of five different lengths (500 mm to 2500 mm) were generated. The four fitted models for $f_{t,0,\mathrm{cell}}$ were used. For each length a total of 10 000 simulations per $f_{t,0,\mathrm{cell}}$ model were performed. The simulation of different lengths does not require any special consideration other than generating the needed amount of cells according to Eqs. (6.2), (6.6) and (6.7). After this, the smallest $f_{t,0,\mathrm{cell}}$ value of each virtual board is taken as the $f_{t,0,\mathrm{glob}}$ value, i.e. $f_{t,0,\mathrm{glob}} = \min(f_{t,0,\mathrm{cell}})$.

Figure 6.10 presents the results for the simulations of the five different lengths, where a clear size effect can be observed for the four studied models. Models (a) and (b) (Weibull and Beta) show a very similar evolution in both mean and spread. More importantly, both models simulate the tensile strength of the reference length ($n = 15 = 1500$ mm) almost exactly on the mean level, and slightly underestimate the standard deviation 6.2 % and 4.5 % for models (a) and (b), respectively. For comparison see results for all boards in Table 5.9.

Model (c), i.e. the Weibull regression model, reveals a similar size effect trend, however, it fails to reproduce the mean and spread for the reference length by 4.9 % and 8.9 %, respectively. A similar behavior is observed for model (d), where the mean differs from the experiments by 6.3 % and the standard distribution by 2.7 %. The reason for this discrepancy is probably related to the specific $E_{t,0,\mathrm{glob}}$ distribution used to generate the needed input data. Owed to the large heterogeneity in the quality of the studied boards and the relative low number of experiments, it is well possible that the statistical distribution obtained for $E_{t,0,\text{glob}}$ does not represent the data with the needed accuracy for this application. This could only be solved with a larger dataset.

In order to quantify the simulated size-effect, a simple exponential function of the form

$$f_{t,i} = f_{t,\text{ref}} \cdot \left(\frac{\ell_{\text{ref}}}{\ell_i}\right)^{\xi} \tag{6.8}$$

was fitted to the results of each of the four models, where $f_{t,i}$ is the tensile strength for a length ℓ_i and $f_{t,\text{ref}}$ is the strength related to the reference length, ℓ_{ref}, 1500 mm. The exponential quantification of the size effect as a generalized description of the size impact, irrespective of underlying failure mechanisms (weak-link concept or fracture mechanics), has become an internationally adopted approach in test evaluation and design codes, e.g. EN 384 (2016), EN 1995-1-1 (2010), ASTM D245 (2019) and ASTM D6570-18a (2018). The obtained results for the exponent ξ are presented in Table 6.4. Figure 6.10 shows exemplarily a graphical representation of the exponential size effect curves obtained for model (b) on the mean and 5% quantile levels for all boards. Considering all boards, the size effect exponents ξ obtained on the mean level range from 0.32 to 0.36, whilst for the 5 % quantile ξ moves between 0.27 and 0.36. For the subset of grades LS10 and LS13 only, the exponents ξ are slightly lower, ranging from 0.30 to 0.33 on the mean level, and between 0.23 and 0.34 for the 5 % quantile. Regarding an assessment of the derived exponents, for oak boards, as well as for any other hardwood species, no reference values could be found. Hence, in a first approach, the quantitative discussion of the simulation-based ξ-values is related to the size factors given in literature for softwoods.

Based on a meta study of an extensive data basis (27 subsets, 3070 specimens) by established Showalter et al. (1987), Lam and Varoglu (1990) and Madsen (1990) on north American softwood—Canadian Spruce-Pine-Fire (SPF) and Southern pine (US)—of different structural grades, Barrett and Fewell (1990) proposed an exponent of $\xi = 0.17$ for the 5 %-quantile level. Further, it was concluded that the length effect in tension and bending is alike. Rouger and Barrett (1995) supported the stated length effect exponent $\xi = 0.17$ and discussed the variables that can influence the size effect of a sample. These are mainly: (i) the sawing pattern and (ii) the used grading system. The sawing pattern may lead to an apparent size

Table 6.4. Computed exponents ξ describing the size effect of the tensile strength for the simulated sets of boards

Model:	(a)	(b)	(c)	(d)
All boards:				
ξ – mean	0.36	0.36	0.32	0.32
ξ – 5%-quant.	0.27	0.28	0.36	0.30
LS10+LS13 grades:				
ξ – mean	0.33	0.33	0.30	–
ξ – 5%-quant.	0.23	0.23	0.34	–

factor up to 0.37. The effect of the grading system was illustrated with lower and higher quality samples graded both visually and with regard to MOE by machine grading. While for the MOE-graded material $\xi = 0.23$ was found, the visually graded boards showed ξ-values ranging from -0.5 to 0.1. Burger and Glos (1996) conducted an experimental study with 750 European spruce (*Picea abies*) boards with knot area ratios ranging from 0.16 to 0.61. The valid ($n = 730$) test results from three significantly different lengths (0.15 m, 1.0 m and 2.5 m) yielded size exponents of 0.13 and 0.22 on the mean and characteristic level, respectively.

Considering the cited literature on softwood results, it can be stated that the simulation-based length effect for the oak boards is roughly two to three times higher when looking at the mean strength level. This holds true for the 5 %-quantile level, too, when considering all investigated oak boards. However, focussing exclusively on the LS10+LS13 subset, being more relevant for structural purposes, the situation is different. The parametric models (a) and (b) yielded for the oak boards coinciding ξ-values of 0.23 which did not differ extremely from the softwood results, whereas the Weibull regression model (c) gave a much higher value, rather close to the mean level results. It is beyond the scope of this work to deepen the discussion of the model-dependent differences. More generally speaking, the differences between the growth-bound macroscopic structure, e.g. of fiber angle and knot features of softwoods and hardwoods—here specifically oak wood—are large enough to justify the assumption, that there might be a marked difference in the respective inherent size effects.

From an overall perspective, it can be concluded that the presented order statistics approach, based on localized (cell-wise) strength values, provides conceivable results for the length effect. The magnitude of the length effect exponent, including the model validation of the 5 %-quantile level, must however be validated experimentally with larger samples.

6.4.4 Correlation of MOE and tensile strength at global level

It was mentioned above that the parametric models (a) and (b), derived for the simulation of $f_{t,0,\mathrm{cel}}$, cannot reproduce a correlation with the MOE at a global level. This is because the global tensile strength, $f_{t,0,\mathrm{glob}}$, results directly from the stochastic process $S_{t,0,\mathrm{cell}}$, while the global MOE is defined a priori for each board according to the corresponding statistical distribution. Vice versa the $f_{t,0,\mathrm{glob}}$–MOE correlation is not a problem for the case of the regression models (c) and (d), since the defined global MOE is considered in the model, thus producing the needed correlation. However, the more simple parametric models, (a) and (b), can be used too, if an additional step is introduced to consider the global correlation $E_{t,0,\mathrm{glob}}$–$f_{t,0,\mathrm{glob}}$.

The method regarded here considers a rearrangement of the $E_{t,0,\mathrm{glob}}$ values assigned to each board, so that the complete sample of $E_{t,0,\mathrm{glob}}$ values correlates with the $f_{t,0,\mathrm{glob}}$ values of each board. In detail, the following steps are followed to simulate the $E_{t,0,\mathrm{cell}}$ and $f_{t,0,\mathrm{cell}}$ profiles:

1. Firstly, a pair of correlated vectors of length N is generated for $E_{t,0,\mathrm{glob}}$ and $f_{t,0,\mathrm{glob}}$, according to Section 6.2. The vectors are called $\underline{\boldsymbol{E}}_{t,0,\mathrm{glob}}$ and $\underline{\boldsymbol{f}}_{t,0,\mathrm{glob}}$, respectively.
2. Then, a total number of N normalized MOE profiles with k segments ($\ell = k \cdot 100\,\mathrm{mm}$), $Z_{t,0,\mathrm{cell}}$, are generated [Eq. (6.1)]. Based on the generated $Z_{t,0,\mathrm{cell}}$ a total of N tensile strength profiles are generated [Eqs. (6.6),(6.7)].
3. The vector $\underline{\boldsymbol{f}}^{\mathrm{cell}}_{t,0,\mathrm{glob}}$ is computed, taking the minimum tensile strength of each simulated board.
4. The following procedure is used to find the indices to reorder the vector $\underline{\boldsymbol{f}}^{\mathrm{cell}}_{t,0,\mathrm{glob}}$, so that it will be correlated to $E_{t,0,\mathrm{glob}}$ in the same manner as $E_{t,0,\mathrm{glob}}$ is correlated to $f_{t,0,\mathrm{glob}}$. In pseudocode the steps are the following:
   (i) $\underline{\boldsymbol{i}} \longleftarrow \mathrm{argsort}(\underline{\boldsymbol{f}}_{t,0,\mathrm{glob}})$
   (ii) $\underline{\boldsymbol{j}} \longleftarrow \mathrm{argsort}(\underline{\boldsymbol{f}}^{\mathrm{cell}}_{t,0,\mathrm{glob}})$
   (iii) $\underline{\boldsymbol{r}} \longleftarrow \mathrm{argsort}(\underline{\boldsymbol{i}})$
   (iv) $\underline{\boldsymbol{q}} \longleftarrow \underline{\boldsymbol{j}}[\underline{\boldsymbol{r}}]$
   (v) $\underline{\boldsymbol{f}}^{\mathrm{cell}}_{t,0,\mathrm{glob}} \longleftarrow \underline{\boldsymbol{f}}^{\mathrm{cell}}_{t,0,\mathrm{glob}}[\underline{\boldsymbol{q}}]$

(Note: the function "argsort()" returns the indices that sort a vector in ascending order, e.g. `argsort([30, 40, 10, 20])` returns the indices `[3, 4, 1, 2]`.)

The steps (i)–(v) should be read as (i) find the indices that sort the vector $\underline{f}_{t,0,\mathrm{glob}}$ in ascending manner and store these indices in the vector $\underline{i}$; (ii) then, the indices that sort $\underline{f}^{\mathrm{cell}}_{t,0,\mathrm{glob}}$ are stored in the vector $\underline{j}$. (iii) The indices that sort $\underline{i}$ are stored in the vector $\underline{r}$; (iv) the vector $\underline{r}$ is then used to re-order the vector $\underline{j}$. (v) If the vector of indices $\underline{q}$ is used to re-order $\underline{f}^{\mathrm{cell}}_{t,0,\mathrm{glob}}$, then $\underline{f}^{\mathrm{cell}}_{t,0,\mathrm{glob}} \approx \underline{f}_{t,0,\mathrm{glob}}$ (see Fig. E.1). Therefore, the indices $\underline{q}$ are used to reorder the list of generated $f_{t,0,\mathrm{cell}}$ profiles, which ensures the correlation between $E_{t,0,\mathrm{glob}}$ and $f_{t,0,\mathrm{glob}}$ for large datasets. The graphical example of Fig. E.1 in Appendix E illustrates this process.

The complete procedure to simulate correlated MOE and $f_{t,0}$ profiles—assuming that the parametric models are used—for a total of N boards with correlated $E_{t,0,\mathrm{glob}}$ and $f_{t,0,\mathrm{glob,cell}}$ variables is depicted as flowchart in Fig. 6.11.

6.5 Localized compressive strength

The variation of axial compressive strength within board, $f_{c,0,\mathrm{cell}}$, was not studied in this thesis. However, for the implementation of a finite element model, a reasonable model for this mechanical property is needed. It was decided to use the same method as described in Section 6.4 for the localized tensile strength. The only difference here is the distribution, $F(x)$, used to represent the compressive strength variation.

Compression strength data is available for datasets B and C, made on GLT specimens of lengths 500 mm and 920 mm, respectively. Compressive failure is very localized. Therefore, applying the same analysis as for the tensile strength, the total length of the specimen can be virtually subdivided into cells of 100 mm in length and then maximize the likelihood function defined in Eq. (2.8) to obtain a rough estimate of the distribution for $f_{c,0,\mathrm{cell}}$ along the board. This method was applied to both datasets B and C with the three-parameter Weibull distribution, and the results are used for the simulations in Chapter 8.

6.6 Discussion

This chapter presented new models for the simulation of mechanical properties within boards, as well as a methodology to preserve a given correlation between the

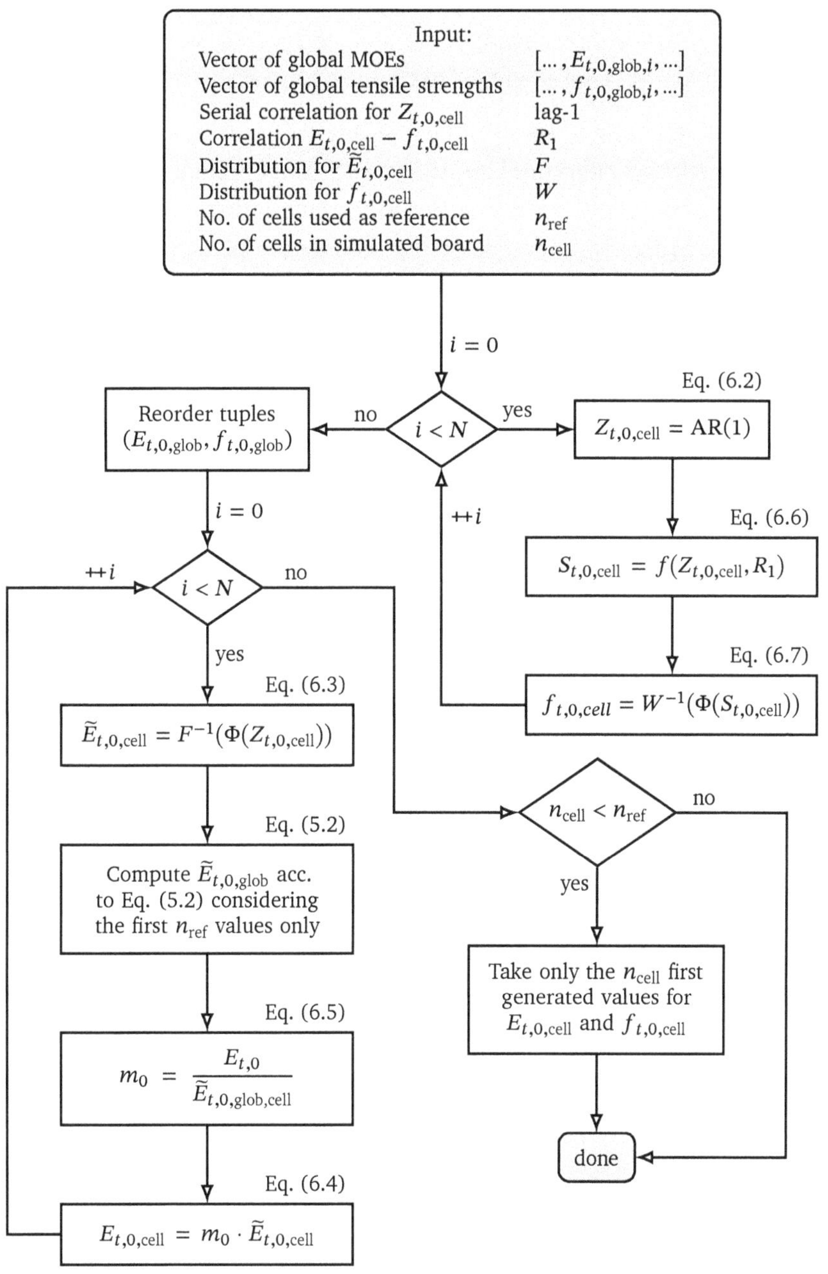

Figure 6.11. Flowchart of the simulation process for $E_{t,0,\text{cell}}$ and $f_{t,0,\text{cell}}$ for the set of all boards, considering the correlation between MOE and tensile strength

global modulus of elasticity, tensile and compressive strength. Within this framework the most important step is the estimation of the distributions for the variation of properties within board (specifically for $\widetilde{E}_{t,0,\mathrm{cell}}$ and $f_{t,0,\mathrm{cell}}$, see Chapter 5), as these will constrain the generated values to realistic ranges.

The detrimental effects on strength of the typically used indicator properties (e.g. KAR or fiber angle), is now captured statistically by the mentioned distributions, rendering the use of the indicator properties in the simulation model redundant. In fact, only one set of data is generated for the simulation of $E_{t,0,\mathrm{cell}}$ for each board [the canonical AR(1) process $Z_{t,0,\mathrm{cell}}$], after which it is successively transformed by different functions (see Fig. 6.3). The application of different functions to the original data adds an extra layer of complexity, however not necessarily making it more complicated, as the individual steps remain relatively simple.

The fact that the simulation model for $f_{t,0,\mathrm{cell}}$ implicitly considers the so-called size effect, makes it very useful for representing the properties of boards with different lengths (as it is the case in glulam beams). For this to work correctly, the right reference length has to be supplied, i.e. the free length used during the tensile tests. Failing in doing so will have a direct effect in the properties simulated for boards with different lengths.

The process whereby the global properties for boards are generated could, in theory, be left out of the simulation, if a regression model is used to define $f_{t,0,\mathrm{cell}}$ and $f_{c,0,\mathrm{cell}}$ (e.g. models (c) and (d) in Table 5.11). This is because during the generation of $f_{t,0,\mathrm{cell}}$ values a correlation with $E_{t,0,\mathrm{glob}}$ is implicitly established, owed to the use of $E_{t,0,\mathrm{glob}}$ as explanatory variable.

Finite element model for the prediction of bending strength in glulam beams

7.1 General remarks

An accurate description of the mechanical properties of glulam beams is essential for a safe and economical design of timber structures. However, owed to the random nature of the mechanical properties of the boards and finger-joints, the estimation of strength and stiffness properties of GLT beams is not trivial and requires the introduction of rather complex models. This means that any attempt to estimate the mechanical behavior of a sample of GLT beams must consider the stochastic component originating in boards and finger-joints. Owed to the intrinsic mechanical aspect of this problem, the finite element method (FEM) presents itself as the natural tool to study the effects of the mentioned variability in GLT beams, as shown in a number of related studies, i.a. Foschi and Barrett (1980), Kandler and Füssl (2017), Fink (2014), and Blaß and Frese (2006).

Within the scope of FEM a few methods have been developed to study stochastic problems , commonly referred to as “stochastic finite element methods” (SFEM), varying in computational cost, ease of implementation and result accuracy for a given problem. A summary of the existing SFEM approaches and their application to GLT beams can be seen in Stefanou (2009) and Kandler et al. (2015a). Among the different SFEM approaches, the Monte Carlo (MC) method gives generally the best results, however at a very high computational expense (Stefanou, 2009). In addition, the MC method is fairly easy to implement, as it only requires the repetition of the simulation with new variable values, representative of the studied problem. For these reasons MC simulations are widely used in stochastic analysis of different fields, and is therefore chosen for this study too.

The present model is similar to—and was inspired by—previous models presented i.a. by Ehlbeck and Colling (1987), Blaß et al. (2005), and Fink (2014) (see Section 2.6), where a GLT beam is subdivided in laminations, and the laminations are subdivided in cells. Material properties are assigned to these cells and a failure criterion is defined to obtain the bending strength. Different from previous approaches, the present model considers fracture mechanics by introducing the extended finite element method (XFEM) to simulate the onset and propagation of *cracks* within the simulated beam. It is expected that a good calibration of the needed parameters (mainly fracture energies) should allow for a good estimation of the bending strength distribution.

In the following, the simulation model—here named "Stuttgart Stochastic Strength Glulam Model (S^{3}GluM)"—will be described in detail, starting with the generation and handling of input parameters, then moving to the implementation of the FE model and post-processing.

7.2 Input parameters

The simulation process begins with the generation of the input data (material properties) according to the models presented in Chapter 6. The objective of this step is to produce a series of virtual boards, resembling the mechanical properties of the boards used in the production of a specific batch of glulam beams. In order to do this, an estimation of the total number of boards needed for the MC simulations is required. For a homogeneous GLT beam, the total number of boards to be generated, n_b, is:

$$n_b = 1.2 \cdot \frac{\ell_{\text{beam}} \cdot n_{\text{lams}} \cdot N_{\text{sim}}}{\overline{\ell_b}}, \tag{7.1}$$

where ℓ_{beam} is the length of the GLT beam, n_{lams} is the number of laminations in the GLT beam ($\ell_{\text{lam}} = \ell_{\text{beam}}$), and $\overline{\ell_b}$ is the mean of the length of the boards. For

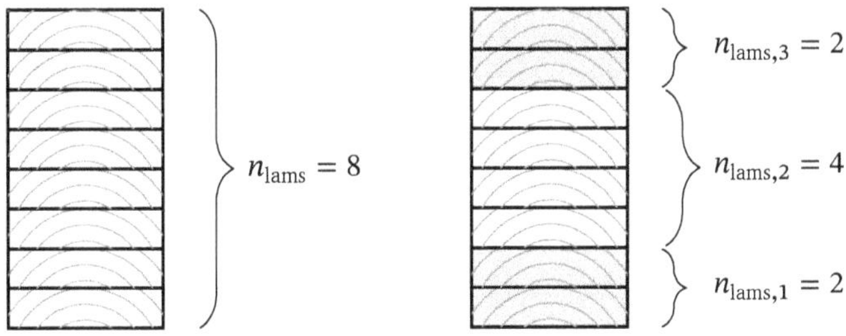

Figure 7.1. Visual explanation of the considered number of laminations per region for the computation of the total number of boards to be simulated

the case of an inhomogeneous build-up Eq. (7.1) is used for each region j, giving a total number of boards to be generated per region, $n_{b,j}$. For this, the total number of laminations per region $n_{lam,j}$, is used (see Fig. 7.1).

The next step consists on simulating a total of $n_{b,j}$ board lengths, ℓ_b, drawn from the corresponding statistical distribution according to Section 7.3.2. Then a set of global mechanical properties $E_{t,0,\text{glob}}$, $f_{t,0,\text{glob}}$ and $f_{c,0,\text{glob}}$ are generated for each board according to their experimental distributions and correlation coefficients, as shown in Chapter 6.2. Thereafter, localized material properties ($E_{t,0,\text{cell}}$, $f_{t,0,\text{cell}}$, $E_{c,0,\text{cell}}$ and $f_{c,0,\text{cell}}$) are generated according to the models presented in Chapter 6.

The generated data for $E_{t,0,\text{cell}}$, $f_{t,0,\text{cell}}$, $E_{c,0,\text{cell}}$ and $f_{c,0,\text{cell}}$ is then stored in an array and saved as a text file with the structure specified in Table 7.1. Each row of this table contains the information of one cell, as well as the associated board and relative position within the board. The structure of the array mirrors the concept of the *endless lamella* used in the production of glulam (see Section 1.3), and helps in the later assignment of material properties to the FE model.

The final step consists of the generation of the mechanical properties for the

Table 7.1. Generated array with the mechanical data along each board

N	No. Board	No. Cell	$E_{t,0,\text{cell}}$	$f_{t,0,\text{cell}}$	$E_{c,0,\text{cell}}$	$f_{c,0,\text{cell}}$
1	1	1	$E_{t,0,1,1}$	$f_{t,0,1,1}$	$E_{c,0,1,1}$	$f_{c,0,1,1}$
2	1	2	$E_{t,0,1,2}$	$f_{t,0,1,2}$	$E_{c,0,1,2}$	$f_{c,0,1,2}$
3	1	3	$E_{t,0,1,3}$	$f_{t,0,1,3}$	$E_{c,0,1,3}$	$f_{c,0,1,3}$
4	1	4	$E_{t,0,1,4}$	$f_{t,0,1,4}$	$E_{c,0,1,4}$	$f_{c,0,1,4}$
5	2	1	$E_{t,0,2,1}$	$f_{t,0,2,1}$	$E_{c,0,2,1}$	$f_{c,0,2,1}$
6	2	2	$E_{t,0,2,2}$	$f_{t,0,2,2}$	$E_{c,0,2,2}$	$f_{c,0,2,2}$
7	2	3	$E_{t,0,2,3}$	$f_{t,0,2,3}$	$E_{c,0,2,3}$	$f_{c,0,2,3}$
⋮	⋮	⋮	⋮	⋮	⋮	⋮
n_{cells}	n_b	j	$E_{t,0,i,j}$	$f_{t,0,i,j}$	$E_{c,0,i,j}$	$f_{c,0,i,j}$

finger-joints. This is done following the procedure described in Section 7.3.2, process from which an array with the structure shown in Table 7.2 is generated.

Table 7.2. Assignment of MOE and tensile strength to the finger-joints of connected pairs of boards

N	Board_i	Board_j	$E_{t,j,i}$	$f_{t,j,i}$
1	1	2	$E_{t,j,1}$	$f_{t,j,1}$
2	2	3	$E_{t,j,2}$	$f_{t,j,2}$
3	3	4	$E_{t,j,3}$	$f_{t,j,3}$
$\vdots$	$\vdots$	$\vdots$	$\vdots$	$\vdots$
$n_b - 1$	$n_b - 1$	n_b	$E_{t,j,n-1}$	$f_{t,j,n-1}$

7.2.1 Configuration files and hashing of parameters

From the models presented in Chapter 6 it is evident that a large number of parameters are involved in the generation of the material properties. The FE model itself adds a considerable amount of extra parameters, too. It is therefore useful to handle all the parameters of a model in a unified manner by means of a *configuration file*. For this, the `json`-format was chosen, which allows for a hierarchical and readable syntax. A `json`-file can be parsed easily by a python script using the standard libraries, and the content of the file can be later validated by means of a *hash* function. This hash function returns a unique alphanumeric sequence (known as "hash") associated with the information contained in the `json`-file. An (simplified) example of such a configuration file is presented in Listing 7.1.

This is especially useful when a parametric analysis is made, where different variables might change. Knowing exactly which parameters lead to a certain result offers great advantages. Without the hash check, it is possible that the user inadvertently sets wrong a parameter, or forgets to change one variable as it was intended. The use of a configuration file, together with a hash check, removes the source of some typical human errors.

7.3 Finite element model

The simulations were performed with the commercial software Abaqus (2017). This program exposes a python API, which allows for the construction of parametric models with the flexibility of an object-oriented programming language. This means that the code can be structured in a comprehensible manner, being a great advantage for complex models. In the following, the different aspects of the

Listing 7.1: Example of a configuration file used in the simulations

```
{
    "name": "Title-of-parameters-set",
    "width": 100.0,
    "t_lam": 20.0,
    "board_1": {
        "ref_length": 1500,
        "length": {
            "s": 0.20,
            "loc": 0.0,
            "scale": 1000.0
        },
        "moe": {
            "distribution": "lognorm",
            "params":{
                "s": 0.370,
                "loc": 6224.42,
                "scale": 5036.21
            }
        },
        "ft": {
            "distribution": "weibull_min",
            "params": {
                "c": 2.211,
                "loc": 5.92,
                "scale": 26.31
            }
        },
        "fc": {
            "distribution": "lognorm",
            "params": {
                "s": 0.0461,
                "loc": -0.226,
                "scale": 50.251
            }
        },
        "ft_fj": {
            "distribution": "lognorm",
            "params": {
                "s": 0.13,
                "loc": 0.0,
                "scale": 40.4
            }
        },
        "moe_norm": {
            "distribution": "loggamma",
            "params": {
                "c": 0.4370,
                "loc":0.9878,
                "scale":0.0481
            }
        },
        "ft_cell": {
            "distribution": "beta",
            "params": {
                "a": 2.1592,
                "b": 1.2302,
                "loc": [6.9325],
                "scale": [93.3778]
            }
        },
        (...) // Additional content that does not fit in the page
    },
    "energies": {
        "glulam": 10.0,
        "fingerjoint": 25.0,
    },
}
```

FE model are presented in detail, starting with the general material properties, geometry and mesh considerations, followed by boundary conditions and failure criteria. A flowchart of the complete simulation process is presented in Fig. 7.2. A description of the FE model and the steps illustrated in Fig. 7.2 are presented in the following.

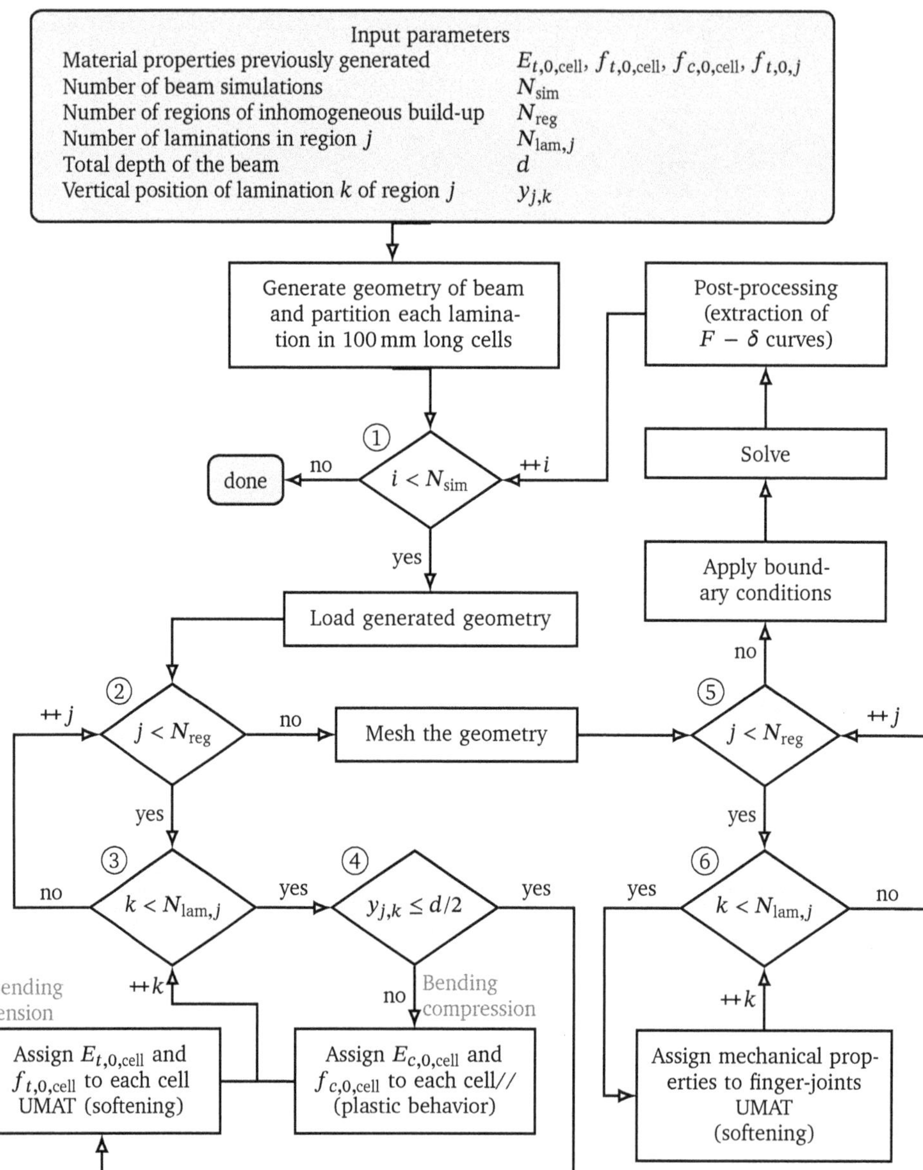

Figure 7.2. Flowchart of the finite element model and Monte-Carlo simulations

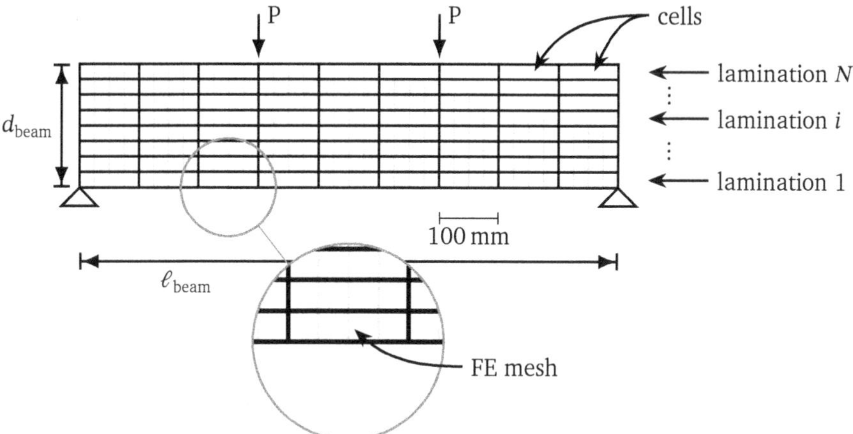

Figure 7.3. Geometry, partitioning and meshing of the finite element model

7.3.1 Geometry and build-up characteristics

Geometry of the finite element model

The geometry of the beams is very simple and can be represented by a rectangle of dimensions length × depth $= \ell_{\text{beam}} \times d_{\text{beam}}$. The partitioning of this geometry is essentially the same as the on used in previous similar models(see Section 2.6), however with two main differences: *(i)* a shorter length of the used cells ($\ell_{\text{cell}} = 100\,\text{mm}$), and *(ii)* the fact that each cell does not correspond to a single FE element, as each cell is further discretized into smaller FE elements.

Specifically, the geometry of the beam is vertically subdivided into the required number n_{lam} of laminations, which are in turn subdivided horizontally into 100 mm long cells (see Fig 7.3). The length (span) of the beams, ℓ_{beam}, chosen according to EN 408 (2012) as $\ell_{\text{beam}} = 18 \cdot d_{\text{beam}}$, does not ensure that ℓ_{beam} is exactly a multiple of 100 mm. Therefore, the length of the cells, ℓ_{cell}, is adjusted so that $\ell_{\text{beam}}/\ell_{\text{cell}}$ is an integer number. Thus, the cell length is defined as $\ell_{cell} = 100\,\text{mm} + \Delta\ell$. For a beam of about 2 m in length, this results in $\Delta\ell = \pm 2\,\text{mm}$ and declines for longer spans, as the amount of cells available to distribute the error increases.

The glulam beams are modeled using two-dimensional, linear plain stress elements with reduced integration (CPS4R). Rigid line elements (R2D2) are applied in the support and loading zones. Orthotropic material behavior was assumed, representing the directions parallel and perpendicular to the grain of the boards.

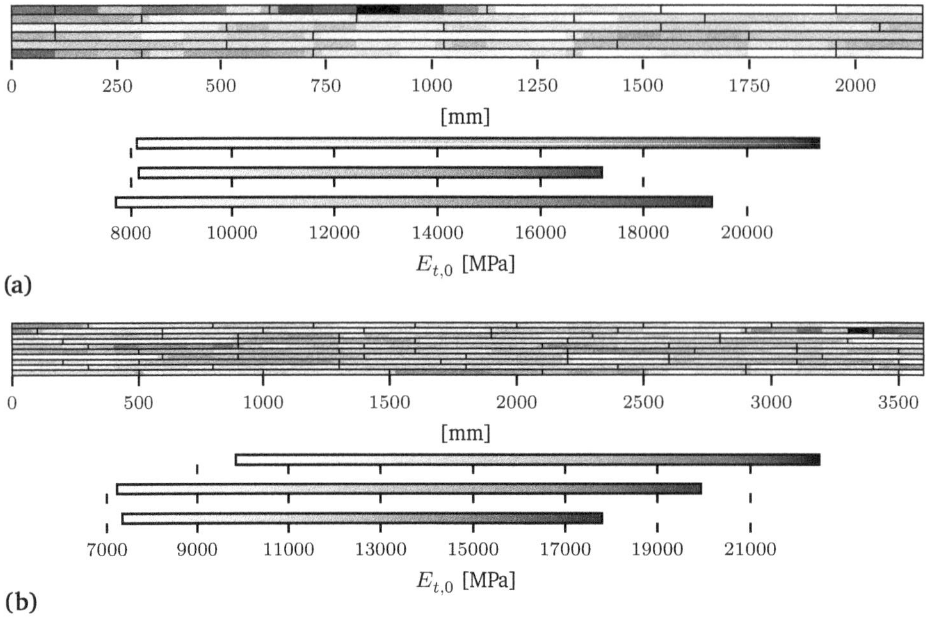

Figure 7.4. Visualization of the material properties distribution within a composite build-up of two different cross-section sizes

Composite build-ups

The model allows for the simulation of so-called composite cross-sections, which, contrasting to homogeneous cross-sections, enable a more efficient use of the material by placing high grade (stiffness/strength) material in the outer regions (higher bending demand), and lower quality boards in the inner zone (see Section 1.3). To consider this, one set of material properties is generated for each region according to Section 7.2. Then, the material assignment process described previously in Section 7.3.2 is used separately in each region. This concept is presented in Fig 7.4a and b, where two composite cross-sections consisting of three regions (LS13-LS10-LS13, according to dataset B) are presented. More precisely, Figs. 7.4a and b show the variation of the simulated $E_{t,0,}$ values throughout two beams with depths of 120 mm and 200 mm, respectively.

7.3.2 Material properties

Length of boards

The simulation of length values for the boards is straightforward and is achieved by sampling values from a continuous distribution and then rounding to the nearest

100 mm. This ensures that the generated lengths are compatible with the 100 mm long cells used in the analyses of Chapter 5. The values used to fit the distribution either corresponded to the measured distances between finger-joints of the tested glulam beams (Dataset B), or to the measured lengths of the boards used to manufacture the GLT (Dataset C).

Tensile strength and MOE of finger-joints

Since no correlation is currently known between any of the mechanical properties of the boards and the tensile strength of finger-joints, $f_{t,0,j}$, no special consideration was made for the simulation of $f_{t,0,j}$. Thus, the values are randomly sampled from the corresponding experimental distribution and assigned to each pair of boards that need to be connected. The finger-joint is considered to have a length equal to the length of one finite element (see Section 7.3.3).

The MOE of each finger-joint, $E_{t,0,j}$, is defined as the average of the $E_{t,0,cell}$ values of the corresponding cells being connected. This means, the MOE of the last cell of one board and the MOE of the first cell of the board to be jointed are used as

$$E_{t,0,j,ik} = \frac{E_{t,0,-1,i} + E_{t,0,1,k}}{2}, \tag{7.2}$$

where “1” and “−1” refers to the first and last cells of a board, and the indices “*i*” and “*k*” identify the boards to be jointed.

Compressive behavior

Plastic behavior was assumed in the zone of bending-induced compression, using isotropic hardening from the predefined material laws present in Abaqus. The model does not consider softening behavior in compression. The plastic behavior follows the curve defined in Fig. 7.5, where plasticity is assumed to start at 75 % of $f_{c,0}$ and the ratio $\varepsilon_u/\varepsilon_p = 3$. A third point was added between the proportional limit and the ultimate strain to produce a curve more similar to the experiments. This curve is based on the experimental results on softwood made by Song et al. (2007).

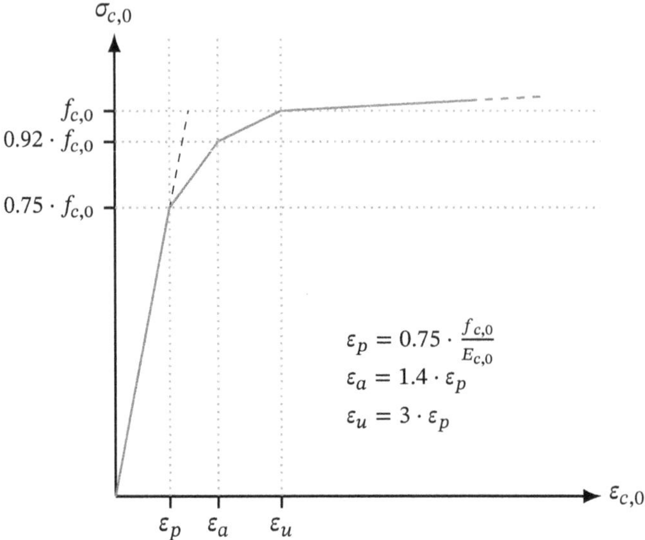

Figure 7.5. Definition of the plastic behavior in the boards of the GLT beam

Tensile behavior

The elements at the bending tension region of the beam were modeled with a user-defined material subroutines (UMAT) written in the Fortran language. Two different user defined materials are implemented: (i) a perfectly brittle material that removes the element once the tensile strength is reached, and (ii) a material with an energy based linear softening as presented by Blank et al. (2017).

This softening law is based on an energy-based stress-displacement response in order to avoid (or minimize) excessive mesh dependencies. The stresses decrease linearly with increasing crack opening, δ, until reaching the failure point δ_f (see Fig. 7.6a). Thereafter the element is removed from the simulation. This behavior is implemented at the material level by means of the consideration of a *virtual* crack which is smeared within the element. The material behavior is shown in Fig. 7.6b, where the stress-strain response can be seen. The evolution of the stress-strain curve of Fig 7.6b can be explained as follows: (i) a linear behavior is observed (Fig. 7.6b) until the tensile strength, $f_{t,0}$, is reached; then (ii) the stresses decline as the opening of the smeared crack, δ, widens (Fig. 7.6e); (iii) during this phase the elastic strains, ε, are reduced at the same rate as the stresses. (iv) The failure process finalizes once the condition $\tilde{\varepsilon} = \tilde{\varepsilon}_f$ is met. It is clear then that the total strain of the element parallel to fiber has two components, and is described as:

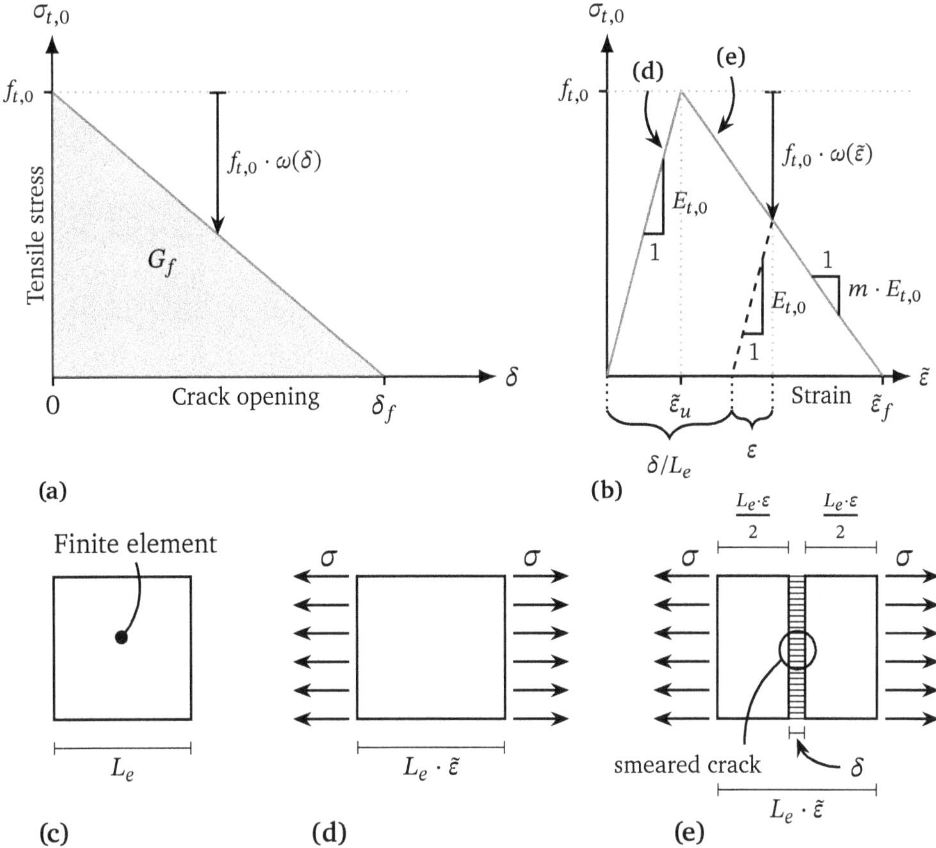

Figure 7.6. Definition of energy-based softening behavior under tensile load according to Blank et al. (2017). (a) Softening as function of crack opening δ; (b) Stress-strain behaviour considering a smeared crack in the element; (c) finite element dimension; (d) finite element under linear-elastic loading ($\tilde{\varepsilon} < \tilde{\varepsilon}_u$); (e) conceptual illustration of finite element during crack growth and softening (no real crack is modelled)

$$\tilde{\varepsilon} = \varepsilon + \frac{\delta}{L_e} \quad , \tag{7.3}$$

where $\tilde{\varepsilon}$ is the total strain, ε is the elastic strain, δ is the opening of the smeared crack in the softening regime, and L_e is the length of the element in the direction parallel to the stresses $\sigma_{t,0}$. Once the softening initiation criteria is met ($\sigma = f_{t,0}$) the stress parallel to the grain is described as follows:

$$\sigma(\tilde{\varepsilon}) = f_{t,0} \cdot (1 - \omega(\tilde{\varepsilon})) \quad , \tag{7.4}$$

where the softening factor $\omega \in [0, 1]$ is defined here as:

$$\omega(\tilde{\varepsilon}) = m \cdot \left(1 - \frac{E_{t,0} \cdot \tilde{\varepsilon}}{f_{t,0}}\right) \quad , \tag{7.5}$$

where $m \cdot E_{t,0}$ is the slope of the softening branch in the strain-stress curve, with m defined as a function of G_f and L_e as:

$$m(G_f, L_e) = \frac{1}{1 - \frac{L_{\mathrm{cr}}}{L_e}} \quad , \tag{7.6}$$

where the critical length, L_{cr}, is defined as:

$$L_{\mathrm{cr}} = \frac{2 \cdot G_f \cdot E_{t,0}}{f_{t,0,}^2} \quad . \tag{7.7}$$

The remaining stress components are likewise reduced by the term $(1 - \omega)$, producing an isotropic softening.

As mentioned by Blank et al. (2017), the softening slope, $m \cdot E_{t,0}$, becomes infinite once the size of the element equals the critical length, L_{cr}. For the cases where $L_e > L_{\mathrm{cr}}$ is satisfied, a positive slope is obtained, which will cause convergence problems due to a snap-back behavior. The length L_{cr} is a function of the fracture energy, MOE and tensile strength, meaning that it varies for the elements of each different cell in the GLT model. In the implementation of the user defined material subroutine (UMAT) the critical length is computed for each integration point. For the case that $L_e \geq L_{\mathrm{cr}}$ is satisfied, a perfectly brittle material behavior is used (i.e. immediate removal of element after reaching $f_{t,0}$). In this manner the stability of the solution is improved.

The UMAT subroutine was tested with a one-element model. The results of this simulation are presented in Appendix C. The code for the UMAT subroutines defining the softening and brittle behavior can be found in Tapia (2022).

Fracture energy

For the consideration of the fracture energy within the GLT model, a distinction is made between wood and finger-joint elements. This entails the assumption that

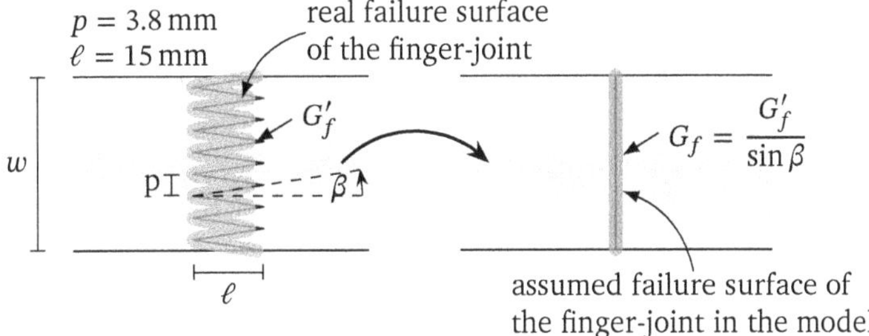

Figure 7.7. Consideration of the geometry simplification of finger-joints for the estimation of G_f

the failure behavior of wood differs from that of the finger-joints. This is reasonable, as the failure of finger-joints not only depends on the material properties of wood, but also on the adhesive used and the geometry of the finger-joint profile (see Fig. 7.7).

The fracture energy of different adhesive types has been studied i.a. by Serrano and Larsen (1999) for Modes I, II and mixed Mode, and by Stapf (2010) for Mode II, for softwoods glued with different adhesives. The values for G_f depend on the type of glue and ranges from $420\,\text{J/m}^2$ (PUR-adhesive) to $1280\,\text{J/m}^2$ (PVAc-adhesive).

The fracture energy of oak wood has been studied i.a. by Aicher (1992), Reiterer et al. (2000) and Reiterer et al. (2002). From these studies a clear conclusion can be drawn, namely that the fracture behavior of oak wood—and in general of hardwoods—tends to be much more brittle than that of softwood species. According to Reiterer et al. (2000) this may be explained by the fact that hardwood cells, like vessel, fiber tracheids and libriform fibers, are much shorter than their equivalents in softwoods.

Since the fracture process can be considered as an aggregation of different macro and microscopic effects, as well as a simplification of the finger-joint geometry (see Fig. 7.7), the direct use of fracture energies obtained experimentally will not necessarily lead to accurate results. Experience in simulations of GLT beams with softening behavior by Blank et al. (2017) has shown that the use of energies of about one order of magnitude higher than those obtained experimentally deliver best results (e.g. Blank et al. used $G_f = 1 \times 10^4\,\text{N/m}^2$). Therefore, in the context of the model for GLT, the fracture energies for finger-joints and wood material can be considered as a variable to be calibrated.

7.3.3 Material assignment and meshing

The material properties are assigned sequentially to each cell, starting at the bottom left corner, and moving to the right end of the lamination (cf. Fig. 7.3). Once the

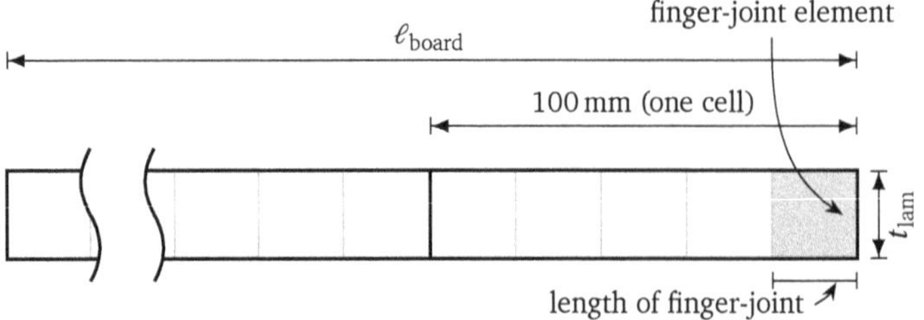

Figure 7.8. Assignment of material properties of the finger-joints

end of the lamination is reached the material properties for the second lamination, starting from the left side, are assigned. For the case of inhomogeneous cross-sections, the different regions have to be considered (see decision elements ② and ⑤ in flowchart of Fig. 7.2). This process is repeated until material properties have been assigned to each lamination (step ③ of flowchart in Fig. 7.2).

The material properties for tension or compression are assigned according to the vertical position of the lamination i (y_i) relative to the mid-depth of the beam ($d/2$). Specifically, tension material properties ($E_{t,0,\text{cell}}$ and $f_{t,0,\text{cell}}$) are assigned if $y_i \leq d/2$, and compression material properties ($E_{c,0,\text{cell}}$ and $f_{c,0,\text{cell}}$) if $y_i > d/2$ (step ④). The shear modulus, G, and MOE perpendicular to the fiber, E_{90}, are scaled according to the values of $E_{0,\text{cell}}$ and the mean values for oak material according to ETA-13/0642 (2013) as:

$$G = G_{\text{mean}} \cdot \frac{E_{0,\text{cell}}}{E_{0,\text{mean}}} \tag{7.8}$$

$$E_{90} = E_{90,\text{mean}} \cdot \frac{E_{0,\text{cell}}}{E_{0,\text{mean}}} \quad . \tag{7.9}$$

The geometry is then meshed with rectangular elements with a side aspect ratio close to 1:1. The edge size of the elements was chosen to be equal to the lamination thickness. Four node, linear plane stress elements with reduced integration (CPS4R) were used.

Finally, the material corresponding to the finger-joints is assigned to the last column of elements of each board, as shown in Fig 7.8 (step ⑥ of flowchart in Fig. 7.2).

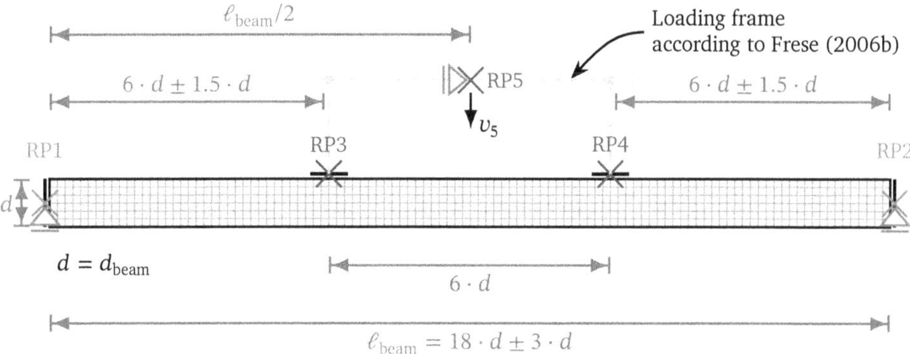

Figure 7.9. Position of reference points (RPs) where the boundary conditions are applied. The loading frame according to Frese (2006b) is shown as reference, too (gray dashed line). Test setup according to EN 408 (2012)

7.3.4 Boundary conditions

The applied boundary conditions (BCs) resemble the test conditions of a four-point bending test according to EN 408 (2012). Due to the pronounced non-linear behavior of the system caused by the softening behavior and plasticity within the model, appropriate BCs are required to improve convergence. This, of course, has to be achieved without sacrificing the correct representation of the experimental boundary conditions.

The loads are applied through rigid lines elements (type R2D2) tied to the top surface of the beam and controlled by reference points RP3 and RP4 (see Fig 7.9). This emulates the steel plates used in the typical experimental setups. For non-linear fracture mechanics analyses a displacement-controlled loading condition is recommended. However, imposing a vertical displacement on each reference point RP3 and RP4 leads to an unbalanced loading condition at both reference points, owed to the stochastic distribution of the mechanical properties (especially $E_{t,0}$) throughout the beam.

Frese (2006b) solved this problem with the incorporation of a frame-like structure connecting both loading points to a *control point* (RP5 in Fig 7.9), which drives the frame. Both loading points are jointed to the frame by means of pinned connections. Now, imposing a displacement v_5 on RP5 produces equal forces at both loading points, conforming to the test conditions according to EN 408 (2012).

This concept can be further simplified by replacing the rigid frame with a linear constraint imposed directly on the vertical displacements of the loading points (RP3, RP4) and the control point (RP5). The linear constraint can be expressed as

$$v_5 - 0.5 \cdot (v_3 + v_4) = 0, \tag{7.10}$$

which can be incorporated in an Abaqus model by means of a constraint of type "Equation". This constraint, similarly as the loading frame, enables a displacement control of the loading process, then producing equal forces at both loading points RP3 and RP4, conforming to the experimental setup described in EN 408 (2012).

Similarly as the loading points, the supports are modelled by rigid line elements. Differently as the experimental setup, the rigid lines are tied to the vertical end faces of the beam, and the respective reference points, RP1 and RP2, are placed at mid-depth (see Fig 7.9). This modification does not have an influence on the bending stresses in the central region of the beam. To improve the convergence, a linear relationship between the horizontal displacements of RP1 and RP2 is considered, too:

$$u_1 + u_2 = 0 \tag{7.11}$$

This forces both supports to move always the exact amount but in opposite directions along the x-axis (horizontal direction). An improvement of the convergence of the system is achieved in this way as sudden, large horizontal movements due to element failure are restrained.

7.3.5 Solver configuration

The model is solved with the standard solver of Abaqus, using the static stress analysis ("StaticStep"), which applies the Newton's method to solve the nonlinear problem iteratively (Abaqus, 2017). A maximum *time* increment of 0.02 was defined, where the time variable moves within $[0, 1]$, and a damping factor of 5×10^{-7} was specified for stabilization purposes. Geometric nonlinearities were not considered.

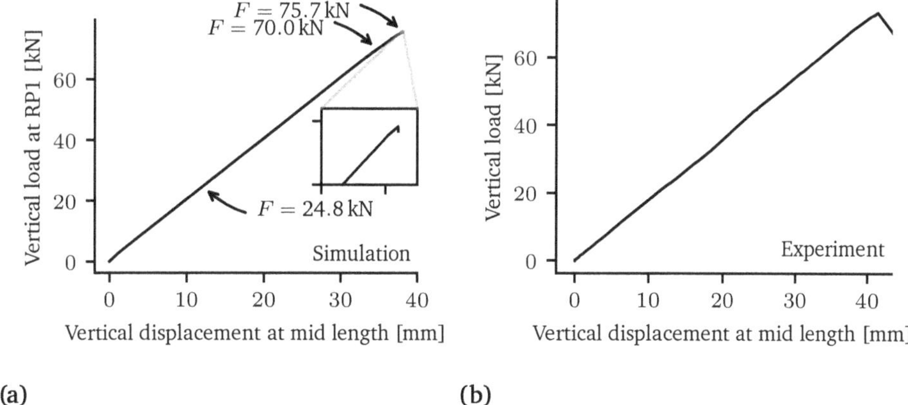

(a) (b)

Figure 7.10. Load-displacement curves corresponding to a cross-section of 200 × 100 mm for (a) a simulated beams; and (b) an experimentally tested beam. The simulation is not intended to replicate the specific experimental result, bur rather show the general similarities between simulations and experiments

7.4 Typical solution and post-processing of results

7.4.1 General failure behavior

A typical simulated load-displacement response is presented in Fig. 7.10a, where the global failure point is reached once the solver cannot find a solution for the next increment due to numerical instabilities. The zoomed-in region in Fig. 7.10a shows the abrupt load drop-off that causes the termination of the simulation. This is similar as observed in some experimental results (see Fig. 7.10b), where mostly rather brittle behaviors are observed. A more detailed description of the behavior of the simulation is presented in Figs. 7.11a–d, where the results in the central region of the simulated beam are shown for two loads, in linear elastic and softening response, respectively (beam properties corresponding to Dataset B). Figure 7.11a shows the location of the finger-joints within the beam. Figure 7.11b shows the bending stresses of the beam in the undamaged linear elastic load regime. This phase ends at around $F = 71$ kN, where the first element reaches its tensile strength, starting the softening behavior.

A further loading of the beam produces increasing damage and redistribution of internal stresses (Fig. 7.11c). The load redistribution is made evident at each step, where increased stress magnitudes are taken by the regions surrounding the element undergoing the softening process. If the lamination above the failing element is strong enough, it will restrain the further development of the crack, increasing the redundancy of the system. At around $F = 75.7$ kN no further load

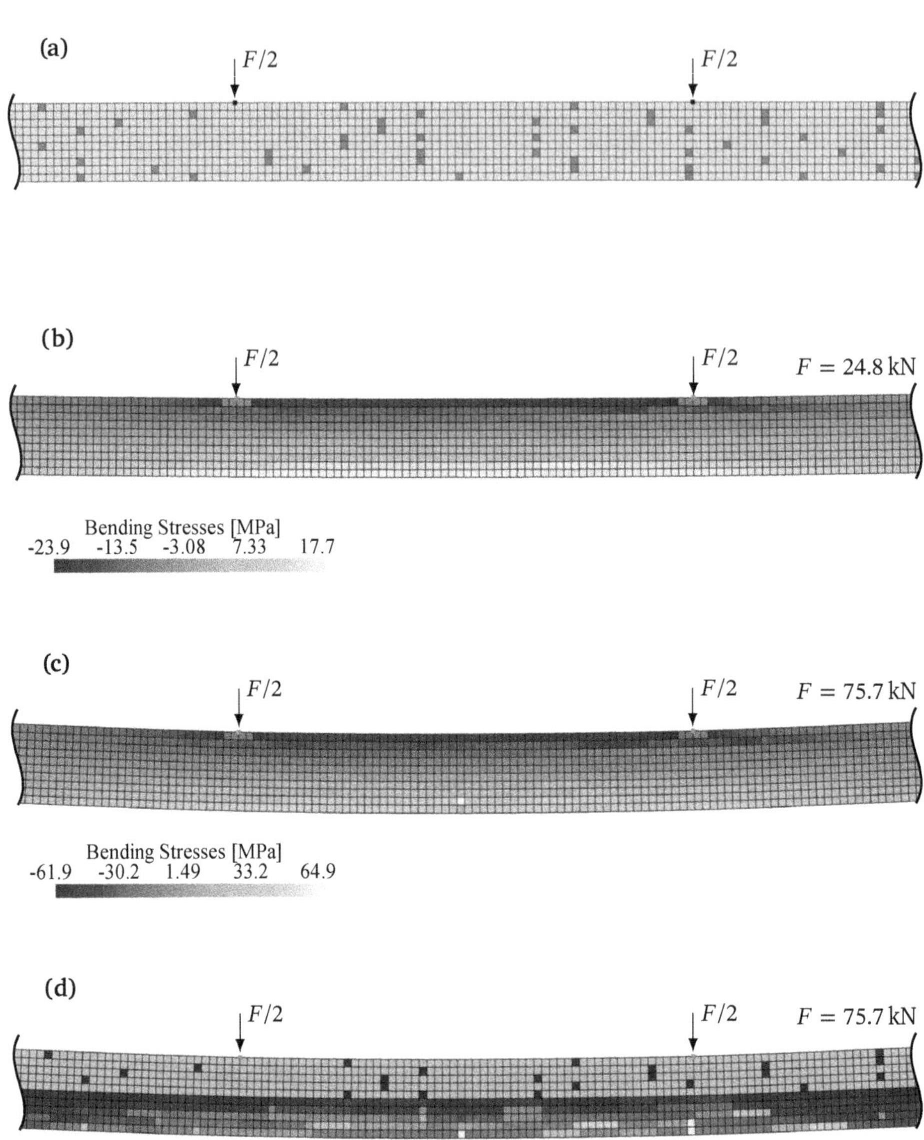

Figure 7.11. Example of a simulated glulam beam. (a) Position of finger-joints. (b) fully elastic, undamaged state, (c) Initiation and progression of damage within the glulam beam, (d) Failure ratio at ultimate load. (Figures created with the Pybaqus library (Tapia, 2021).)

can be taken by the model. This is owed to two finger-joints that reach their respective tensile strengths almost simultaneously (Fig. 7.11d), leading to a load-drop in the last increment, followed by the solver aborting the simulation due to numerical instabilities.

From Figs. 7.11b–d it is evident that the *cracks* (damaged regions) propagate generally in vertical direction (perpendicular to the beam's main axis). Although this constitutes a simplification of the real, experimentally observed failure behavior—where shear stresses and fractures pronouncedly interact after the first failure—the results obtained are in good agreement with the experiments, as shown below. The simplified model assumptions whereby only the tensile strength of boards and finger-joints is considered to asses the failure of the beam are considered, has been made by previous similar models, too. A *realistic* crack propagation could only be achieved by considering shear strengths and fiber deviation too, which most probably will further affect the convergence of such highly non-linear models. Steps in this direction have been taken recently e.g. by Lukacevic et al. (2018), however only at a board level. Applying a stochastic approach with such a model at the GLT level would be prohibitively expensive from a computational point of view.

7.4.2 Data extraction from the FE model

After each simulation the following data are extracted from the model: vertical forces at the five reference points RP1 through RP5, as well as the vertical displacements at RP3, RP4 and RP5. The vertical forces at RP3 and RP4 were used to corroborate that the same force is being applied at both points throughout the simulation [i.e. that Eq. (7.10) had the intended effect]. These forces are also used to compute the load-displacement curves, later used for the determination of the bending strength of each beam.

7.4.3 Definition of bending failure

Although the implemented FE model considers fracture mechanics, meaning that at some point a maximum load is reached, some additional considerations have to be made to obtain sensible results. From experiments on GLT made of oak it was observed that a certain degree of non-critical damage is normally present prior to the global failure of the beam. This is noticeable as slight changes in the inclination of the load-displacement curve, which for Fig. 7.10a begins at around 70 kN. However, the overall response is rather brittle. Similar behavior was observed in the simulation results (see Fig. 7.10b). The mechanics of the fracture process in the model is a simplification and differs from the typically observed cracks that follow the fiber direction of the timber. In the model the cracks propagate vertically and can be effectively restrained by boards with very high tensile strength above them. The bending strength, f_b, is then computed according to linear elastic Bernoulli beam theory as

$$f_b = \frac{3 \cdot F_{\max} \cdot \ell_a}{d^2 \cdot w}, \tag{7.12}$$

where d and w are the depth and width of the cross-section, respectively, and ℓ_a is the distance from the support to the nearest loading point (see Fig. 7.9).

This analysis is performed for all the configurations by means of a python script after all the simulations are done. Within this process the statistics and graphics are produced as well.

7.5 Discussion

The presented FE model is in many aspects very similar to *established* strength models, such as the *Karlsruher Rechenmodel*. In essence the beam geometry is subdivided in laminations, boards, cells and finger-joints, and the material properties are assigned accordingly. Non-linear fracture mechanics (NLFM) was included by means of a smeared crack approach. More *stable* boundary conditions were defined to compensate the strong nonlinearities arising from the damage process. The consideration of NLFM adds complexity to the model in the form of convergence issues and an additional input parameter, namely the fracture energy. The implication of this is that a calibration of the model is needed to obtain the correct parameter for the fracture energy. This will be addressed in the next chapter by means of parametric analyses.

The use of a configuration file to specify all the needed parameters is of great help when analyzing different configurations. This also improves the reproducibility of the results, as none of the relevant parameters is hidden somewhere in the code. Furthermore, although not discussed in depth here, special attention was put into writing the model in an organized manner, using the full potential of the object oriented programming paradigm, in this case implemented in python. This helps achieving a cleaner code and an easier detection of errors. The code is made public and can be found on Tapia (2022).

Model calibration and parametric analysis

8.1 General remarks

Most of the input parameters required by the model, presented in the previous chapter, can be experimentally determined, as shown in Chapter 5 and 6. The fracture energy values are, however, the exception, as experimental values derived from typical small clear wood samples are not expected to deliver good results in this model as pointed out by Blank et al. (2017). This is owed to the fact that the damage process is considered in a smeared manner, simplifying the geometry and failure characteristics along the evolving crack or damaged region. Therefore, adequate values for the fracture energies of finger-joints, $G_{f,\text{fj}}$, and boards, $G_{f,b}$, need to be calibrated with experimental ultimate load data of GLT beams.

The first of this chapter is dedicated to applying the GLT strength model to the datasets B and C, presented in Chapter 3. For this, the exact material properties and geometry used for the simulation of the beams belonging to both datasets are described. Then, the results for both datasets obtained within the linear-elastic regime are discussed. Specifically, the simulated and experimental stiffness (or global MOE) of the beams is compared. The nonlinear fracture behavior is then closely investigated. Here, appropriate values for $G_{f,b}$ and $G_{f,\text{fj}}$ for the datasets B and C are obtained by means of a maximum likelihood approach. In the second part of the chapter, the previous results are extended by means of a parametric analysis, where the board length, cross-sectional depth and material properties are analyzed.

8.2 Studied configurations

8.2.1 Dataset B

Material properties

For the generation of the required profiles of mechanical properties, the models presented in Chapter 6 were fitted with the dataset B (see Section 3.4.2). This dataset does not contain information regarding the variation of properties within boards, hence, some assumptions were made based on previously gained knowledge. Since the boards to be simulated are rather short (≈ 600 mm in length) and large defects were removed, the parameters obtained for *clear wood* sections were used for $\widetilde{E}_{t,0,\mathrm{cell}}$ and $Z_{t,0,0\mathrm{cell}}$. For the variation of tensile strength, the censored Beta model (b) was fitted considering a total of nine cells of 100 mm in each board, from which only one value is known. The reason to choose nine cells is owed to the fact that the standard length for tensile strength, ℓ_t, is equal to nine times the width of the cross-section, w, thus for this case $\ell_t = 9 \cdot w = 9 \cdot 100\,\mathrm{mm} = 900\,\mathrm{mm}$.

The modulus of elasticity under compression parallel to fiber, $E_{c,0}$, is assumed to be equal to $E_{t,0}$. This is a simplification, as the behavior of timber boards is somewhat different in compression as it is in tension. However, since no $E_{c,0}$ data are available for Dataset B, $E_{t,0}$ should hold as a fairly good approximation. An explicit consideration of $E_{c,0}$ is, nevertheless, possible as long as the correlation matrix is available, as the method can be used with an arbitrary number of variables. All the parameters used for the simulation are presented in Tables D.1 and D.2.

Geometry and load setup

Three cross-sections were simulated, corresponding to the three experimental configurations width × depth: 100×120 mm, 100×200 mm and 100×300 mm, representing inhomogeneous build-ups as shown in Fig. 3.2. The length of the beams was $\ell_b = 15 \cdot d$, and the central region, subjected to pure bending, had a length of $6 \cdot d$, complying with the experimental setup and EN 408 (2012).

8.2.2 Dataset C

Material properties

The GLT beams from dataset B were produced with rather long oak boards ($\approx 2\,\mathrm{m}$), classified as D24 according to NF B 52-001-1 (2011). Experimental data for finger-joints were measured from flat-wise bending tests, meaning that a conversion factor need to be used to obtain the tensile strength information required by the FE model. Based on own tests for the used material not reported here, a conversion factor of 0.8 was used, i.e. $f_{t,0,\mathrm{fj}} = 0.8 \cdot f_{m,\mathrm{fj}}$. This value is larger than that obtained by Aicher and Stapf (2014), where an average factor of 0 7 was obtained for datasets of three different producers of oak GLT.

The solid wood boards were tested in both flat-wise bending and tension tests. The tension tests were performed on a different batch as the bending tests, meaning that their properties might differ from the original batch. Nevertheless, since both batches were classified as D24 and the origin was the same, the mechanical properties should be rather similar. Thus, in order to avoid transformation factors, the tension data was considered here.

It was observed that the distribution of $f_{t,0}$ of the boards was very similar to the group of LS10-LS13 boards investigated in Section 5.5. Therefore, the censored Beta model (b) calibrated in Section 5.5 was chosen to model the variation of tensile strength within board, $f_{t,0,\mathrm{cell}}$.

The parameters for the MOE-relevant processes $\widetilde{E}_{t,0,\mathrm{cell}}$ and $Z_{t,0,\mathrm{cell}}$ were taken as those corresponding to the set containing all investigated cells. The global MOEs of the boards were obtained from dynamic excitation measurements, which however tend to deliver between 5 % and 10 % higher values as compared to static MOE measurements.

Geometry and load setup

Two cross-sections were simulated, corresponding to the two experimental configurations (width × depth): 160 × 160 mm and 160 × 300 mm, both representing homogeneous build-ups. The span of the simply supported beams conformed to $\ell_b = 18 \cdot d$, with the constant-moment region having a length of $6 \cdot d$.

8.3 Analysis of stiffness results

8.3.1 Stiffness results of dataset B

Figure 8.1 and Table 8.1 present the computed global MOE for the three studied cross-sections. It is immediately evident that, while a good agreement is observed for the two larger cross-sections of 200 × 100 mm and 300 × 100 mm, a considerable disagreement of about 13 % is obtained for the smallest cross-section of 120 × 100 mm. This is certainly not an expected result, since the MOE of the different cross-sections should, in average, remain very similar owed to the fact that the same material is used (see e.g. Fink, 2014). Bearing this in mind, it can be argued that the simulated results are reasonable, and that either accidentally a sample of beams with boards of significantly lower MOE values were produced—which admittedly is rather unlikely, but possible. Or a problem occurred with the calibration of the LVDT used to measure the vertical displacement, the latter being more likely.

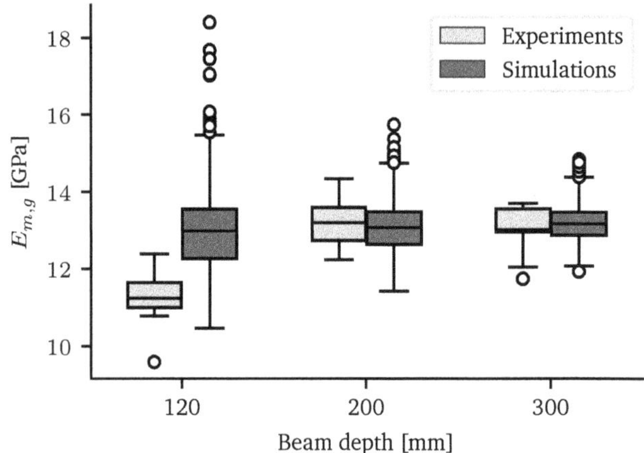

Figure 8.1. Empirical and simulated modulus of elasticity, $E_{m,g}$, for the three analyzed beam depths

In the simulations, the standard deviation of $E_{m,g}$ decays with larger cross-sections, which is expected, as the homogenization effect increases with an increase of used material. Contrary, in the experiments no reduction of the variation for the larger cross-sections was observed, which can be related to the relative low number of specimens tested. Summarizing, it can be stated that the model gives satisfactory stiffness results.

Table 8.1. Experimental and simulation results for the global modulus of elasticity, $E_{m,g}$, of dataset B

Depth [mm]		N [–]	mean [GPa]	std. [GPa]	COV [%]	$E_{m,k}$ [GPa]
120	Exp.	10	11.4	0.9	8	9.5
	Sim.	1000	13.0	1.0	8	11.5
200	Exp.	10	13.2	0.7	5	11.8
	Sim.	1000	13.1	0.6	5	12.1
300	Exp.	10	12.9	0.7	5	11.5
	Sim.	1000	13.2	0.4	3	12.5

Table 8.2. Experimental and simulation results for the modulus of elasticity, $E_{m,g}$, of dataset C

Depth [mm]		N [–]	mean [GPa]	std. [GPa]	COV [%]	$E_{m,k}$ [GPa]
160	Exp.	20	11.2	0.9	8	9.6
	Sim.	1000	11.4	0.8	7	10.1
300	Exp.	20	10.9	0.4	4	9.9
	Sim.	1000	11.5	0.5	5	10.6

8.3.2 Stiffness results of dataset C

The stiffness results can be seen in Table 8.2. A rather good agreement can be observed for the smaller beams. However, for the larger beams the simulations present a mean value about 6 % higher than the experimental results. The coefficients of variation for experiments and simulations are in good agreement.

8.4 Calibration of fracture energy

The FE model is calibrated using two datasets, B and C (see Sections 3.3 and 3.4), respectively. For this, simulations with a range of fracture energy values are made. The results obtained for each G_f value are compared on the basis of their likelihood values, which is computed against the experimental statistical distribution. In this manner, it is possible to (i) determine which fracture energy value better fits the experimental results, and (ii) explain to what degree the consideration of a material softening for the finger-joints improves the obtained solution.

8.4.1 Calibration of fracture energies with dataset B

Fracture energies

The fitting of the unknown fracture energies, $G_{f,\text{fj}}$ and $G_{f,b}$, was performed in two consecutive steps:

1. Firstly, the $G_{f,\text{fj}}$ value is varied between 5 N/mm and 20 N/mm in increments of 1 N/mm, while $G_{f,b}$ is set as zero (brittle behavior). A total of 1000 simulations are performed for each configuration, and the likelihood of each set (i.e. for each depth) is computed against the corresponding experimental distribution.
2. Then, the $G_{f,\text{fj}}$ value that maximizes the likelihood is chosen and new simulations are performed, now varying $G_{f,b}$. The $G_{f,b}$ value that produces the best fit to the experimental results is chosen.

Fracture energy calibration results

The computed log-likelihood values for the analyzed range of $G_{f,\text{fj}}$ are presented in Fig. 8.2a–c for the three experimentally tested beam depths of 120 mm, 200 mm and 300 mm, respectively. A clear maximum can be observed at $G_{f,\text{fj}} = 12\,\text{N/mm}$, $G_{f,\text{fj}} = 11\,\text{N/mm}$ and $G_f = 10\,\text{N/mm}$ for the cross-sectional depths of $d = 120\,\text{mm}$, $d = 200\,\text{mm}$ and $d = 300\,\text{mm}$, respectively. The obtained results can be regarded as consistent for the three different cross-sectional depths, as very similar values of $G_{f,\text{fj}}$ were found to maximize the $\log \mathcal{L}$. A large discrepancy between the optimal $G_{f,\text{fj}}$ values would mean that the model is not suited for the simulation of $f_{m,g}$.

In view of the results presented in Figs. 8.2a–c, a value for the fracture energy for finger-joints of $G_{f,\text{fj}} = 11\,\text{N/mm}$ was chosen as the optimum value for Dataset B. Thereafter, a series of $G_{f,b}$ values were tested, where the best results were delivered by $G_{f,b} = 15\,\text{N/mm}$. It is worth mentioning that the effect of $G_{f,b}$ is much smaller as compared to $G_{f,\text{fj}}$, mostly controlling the lower tail of the $f_{m,g}$ distribution. This is owed to the fact that, for this case, the finger-joints represent the weak regions in the beam, implying that $G_{f,\text{fj}}$ has a larger influence in the failure mechanism as $G_{f,b}$.

Figure 8.3 presents the CDF curves for both, experimental and simulation results ($G_{f,\text{fj}} = 11\,\text{N/mm}$, $G_{f,b} = 15\,\text{N/mm}$) for the three studied cross-sections. The relevant statistics are presented in Table 8.3. A good agreement can be observed between experimental and simulation results, confirming the suitability of the optimized values for $G_{f,\text{fj}}$ and $G_{f,b}$.

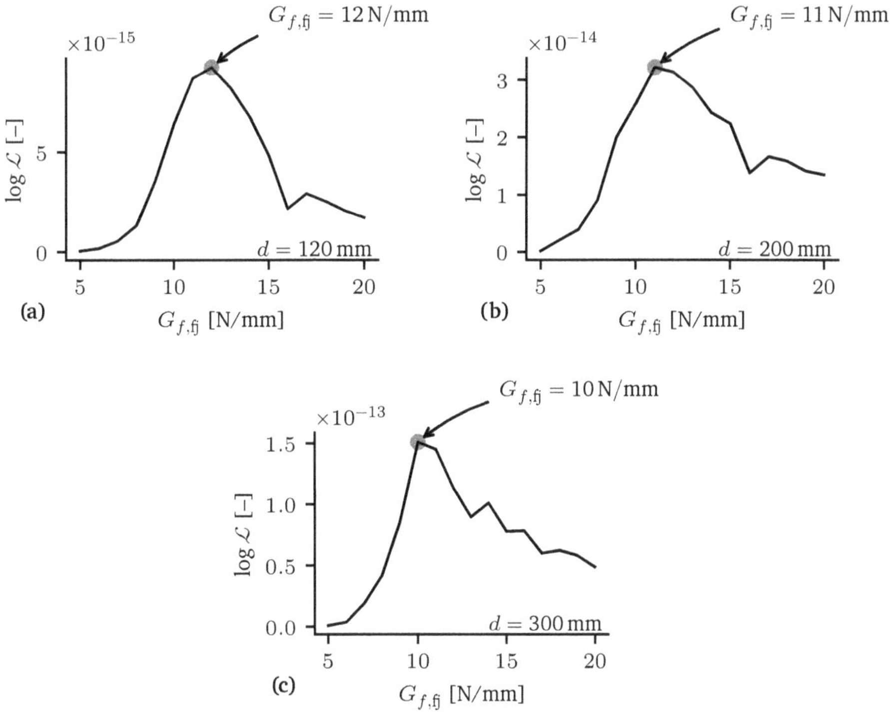

Figure 8.2. Computed log-likelihood for a range of fracture energies for the finger-joints for: (a) $d = 120\,\text{mm}$; (b) $d = 200\,\text{mm}$; and (c) $d = 300\,\text{mm}$

Table 8.3 reveals that the mean bending strength value of the simulations is in average less than 1 % lower than the experimental values. The characteristic values, $f_{m,k}$, are less than 3 % lower for the simulations. Thus, simulation results are slightly on the safe side, which might be regarded as a desired characteristic for such a model.

Discrepancies are seen mainly in the upper tails of the distributions, where the simulations tend to overestimate the empirical results. However, for practical applications the mean and lower levels (especially the 5 % quantile) have a much higher relevance. For the case of the smallest cross-section ($d = 120\,\text{mm}$) the lower tail seems to be underestimated by the simulations. This might be owed to the fact that for shorter beams there are less weak segments within the constant bending region ($= 6 \cdot d$), meaning that a larger number of tests is needed in order to get results with a similar confidence level as the larger cross-sections. The fact that for the two larger cross-sections a good agreement is obtained in the lower tails, supports this hypothesis.

The coefficients of variation are 30 % to 40 % higher for the simulations as

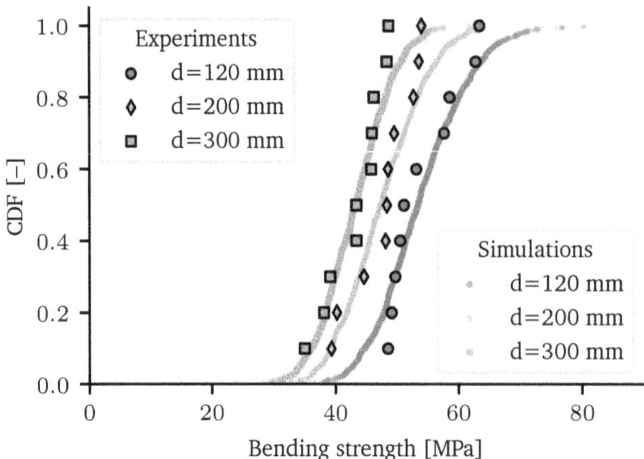

Figure 8.3. CDF of the bending strength for the three cross-sections obtained experimentally and by means of the here presented simulation model

Table 8.3. Statistics for the simulations and experiments with different cross-sections

Depth [mm]		N [–]	mean [MPa]	std. [MPa]	COV [%]	$f_{m,k}$ [MPa]
120	Exp.	10	54.4	5.4	10	43.7
	Sim.	1000	53.8	7.3	14	42.6
200	Exp.	10	47.9	4.9	10	37.7
	Sim.	1000	47.6	6.5	14	37.3
300	Exp.	10	43.4	4.3	10	34.3
	Sim.	1000	43.3	5.7	13	34.2

compared to the empirical results. This difference is most probably due to the mentioned higher values on the upper tails.

Finally, the size effect exponent, ξ, was computed for both, experiments and simulations [see Eq. (6.8)], and is presented in Table 8.4 for the mean and 5 %-quantile levels. A good agreement can be observed between empiric and simulated values, both proving a rather high size effect for the regarded GLT. Compared to the size effect assumed in EN 1995-1-1 (2010) of $\xi = 0.1$, the values obtained here—both numerical and experimental—are clearly much larger. For this particular case the reason can be related to the characteristics of the weak regions of the beam, defined mostly by the finger-joints. The oak material also influences this behavior, which can be assumed from the rather high size effect values obtained in Section 6.4.3. The effect of finger-joints and length of the boards is analyzed in more detail later in this chapter.

Table 8.4. Size effect exponents computed for the experimental and simulation results for dataset B

	mean	5%-quant.
Experiments	0.25	0.26
Simulations	0.24	0.23

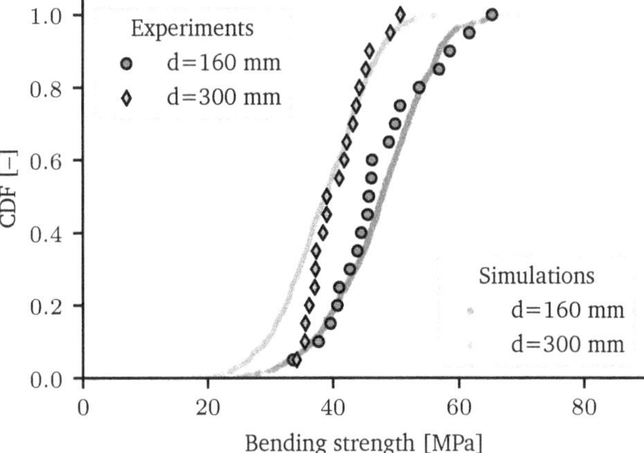

Figure 8.4. Empirical and simulated CDF curves of the two studied cross-sections for dataset C

8.4.2 Application of fitted energy values to dataset C

In this section, the calibrated model is used to simulate GLT beams corresponding to dataset C, which contains experimental data for two cross-sections.

Bending strength results

Figure 8.4 presents the empirical CDF for the bending strength of experiments and simulations, for both analyzed cross-sections. It can be seen that, in general terms, the empirical and simulated curves agree rather well, the main discrepancy being observed in the lower tail of the larger cross-section. There, the simulations considerably underestimate the experimental values, which can be seen in Table 8.5, which contains the relevant statistics. While the simulation values for the smaller cross-section agree with the empirical value very well, the values for the larger cross-section exhibit a difference of 6 % and 18 % for the mean and 5 %-quantile values, respectively.

The observed difference is most likely related to the rather large uncertainty associated to the input data. Specifically, the use of a conversion factor to obtain $f_{t,\text{fj}}$ from bending tests delivers probably a rather vague estimation of the true (unknown) $f_{t,\text{fj}}$ distribution. This applies similarly to $f_{t,0}$, as the tensile strength

Table 8.5. Statistics for the simulations and experiments with different cross-sections

Depth [mm]		N [-]	mean [MPa]	std. [MPa]	COV [%]	$f_{m,k}$ [MPa]
160	Exp.	20	47.6	8.0	17	33.9
	Sim.	1000	47.3	8.0	17	34.2
300	Exp.	20	40.8	4.5	11	32.7
	Sim.	1000	38.5	7.1	18	26.9

was tested in a different batch, potentially presenting some differences with the original material. This was also the reason to not apply the same procedure to fit the fracture energies as done with dataset B.

8.5 Parametric analysis

It is evident that the stiffness and bending strength behavior of the simulated (and the real) beams depend on a large number of parameters. Specifically, it is clear that the distributions of strength and MOE affect the global behavior of the beams in very unique manners. However, an exact analytical quantification of each effect remains a demanding problem. The presented model, with its parametric and stochastic nature, allows to study the effects of different factors on stiffness and bending strength, enabling a better understanding of the interactions of different parameters.

In the following, a parametric analysis is performed in order to study the effects of localized tensile strength models, length of boards and finger-joint quality on the bending strength of simulated GLT beams. In this campaign, the fracture energy for finger-joints is chosen as $G_{f,\text{fj}} = 11\,\text{N/mm}$ and for boards as $G_{f,b} = 15\,\text{N/mm}$, as calibrated above. The rest of the parameters are described in the section corresponding to each analysis.

8.5.1 Effect of tensile strength distribution

In Chapter 5 a total of four different models were derived for the description of the localized variation of tensile strength within board. Later, in Chapter 6, the four models were used to simulate boards of different lengths in order to study the size effect at the level of the individual boards. In this section the same four models are used to generate $f_{t,0,\text{cell}}$ profiles for the simulation of GLT beams under four point bending condition. The objective is to study the sensitivity of the global bending strength of GLT beams with respect to the chosen model for $f_{t,0,\text{cell}}$. For this,

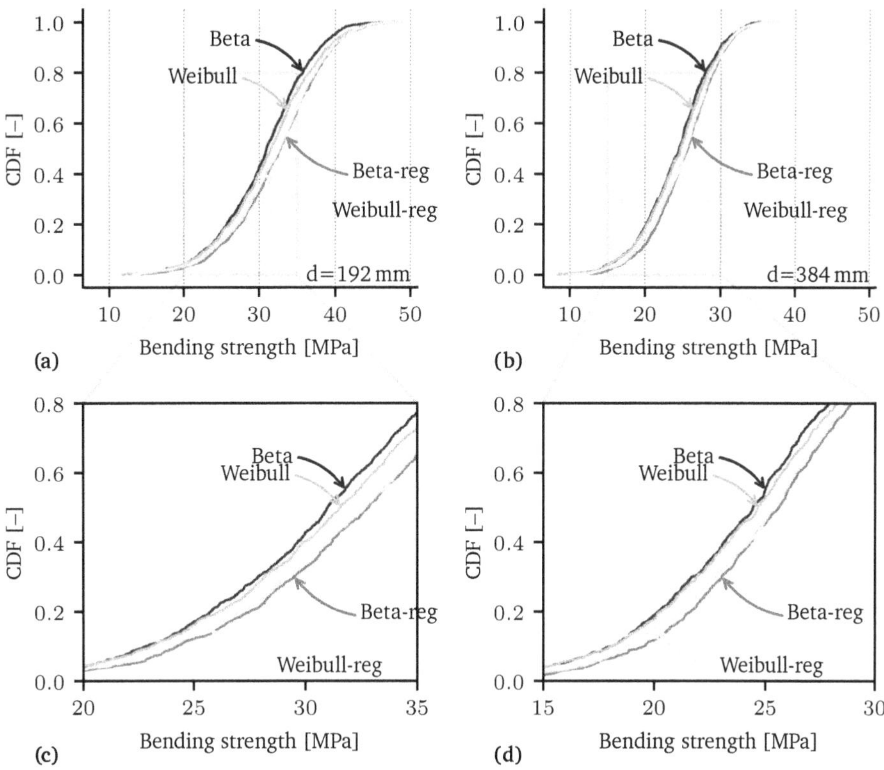

Figure 8.5. CDF curves for the simulated $f_{m,g}$ values for different models of $f_{t,0,\text{cell}}$. (a,c) cross-section of 192 × 175 mm, (b,d) cross-section of 384 × 175 mm

two different depths built-ups of 8 and 16 laminations, respectively, are studied. The tensile strength of the finger-joints correspond to those of dataset B. The parameters for $E_{t,0,\text{glob}}$, $\widetilde{E}_{t,0,\text{cell}}$ and $Z_{t,0,\text{cell}}$ correspond to those of all analyzed cells (see Tables 5.7 and 5.8). The length of the boards is chosen constant and equal to 1000 mm. A total of 1000 simulations are performed for each configuration.

Results

Figures 8.5a,b present the CDF curves of the simulated GLT beams with the four different $f_{t,0,\text{cell}}$ models. It can be seen that there are only small differences between the parametric Weibull (a) and Beta (b) models, and between the regression models Weibull (c) and Beta (d), respectively. There is, however, a clear difference between parametric and regression models, mirroring what has already been observed for the individual boards (see Section 6.4.3).

There is only a minor difference between parametric models (a) and (b) in the upper half of the distribution. This is probably related to the differences in the upper tails of the corresponding models, where the Beta model (b) is capped by a maximum value. Nevertheless, the exact shape of the upper tail of the distribution for $f_{t,0,\text{cell}}$ has a low practical relevance, as for design purposes the most important aspect is to correctly describe the lower tail of the distribution. This is ensured by the chosen method to fit $f_{t,0,\text{cell}}$ by means of survival analysis, as shown in Section 5.5.4. As long as reasonable models are used—i.e. distributions that can describe the lower tail correctly—the sensitivity of the GLT mode to different $f_{t,0,\text{cell}}$ models is very small.

8.5.2 Effect of length of boards

The length of the boards is related to two effects: (i) as the boards get longer, fewer finger-joints are present in the critical loaded region of the beam, while simultaneously, (ii) the probability of encountering a weak region in the board increases. If the finger-joints are of poor quality, then reducing their number seems convenient. However, longer boards translate in a higher probability of encountering a weaker region, which is probably the very reason to having short boards in the first place, as it was in the case of dataset B. It is evident then, that there is an interaction between tensile strength of boards and finger-joints, influenced by the average length of the boards.

To study this effect, lengths of boards between 500 mm and 3500 mm in steps of 500 mm are studied for three cross-sectional beam depths: 120 mm (6 laminations), 240 mm (12 laminations) and 360 mm (18 laminations). The parameters for the different material properties were taken equal as for the beams of dataset C.

Relationship between strength of finger-joints and boards

Before analyzing the results of the parametric analysis, the relation between the tensile strength of boards and finger-joints, dependent on the length of the boards, is discussed. For this, extreme value theory comes in handy. In Section 5.5 it was shown that starting from the cell-wise tensile strength distribution, $F(x)$, the distribution at board (global) level, $F^{(n)}_{\min}(x)$, can be obtained by applying Eq. (2.6) for, in that case, a board composed of $n = 15$ cells, i.e. 1500 mm in length. Following this theory, distributions for $f_{t,0}$ can be obtained for different lengths. This is illustrated in Figs. 8.6a,b for $\ell = 500$ mm and $\ell = 1500$ mm, respectively. The dashed line represents the tensile strength of finger-joints, $f_{t,0,\text{fj}}$, which is assumed equal for both cases, as the length of the board should not affect their resistance.

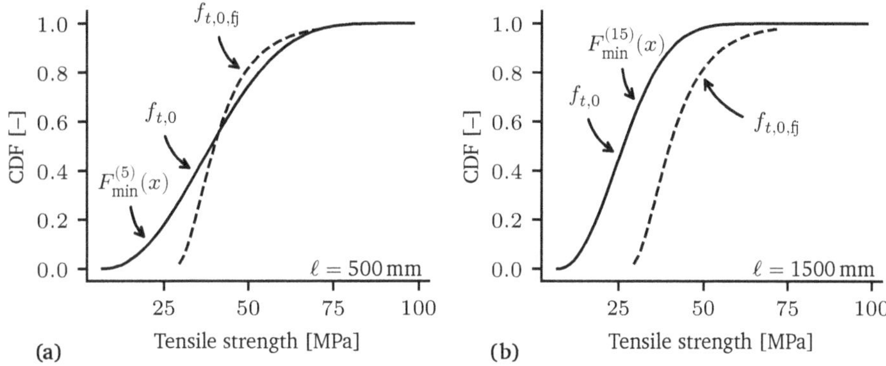

Figure 8.6. Application of extreme value theory to $f_{t,0,\mathrm{cell}}$ for boards of (a) 500 mm and (b) 1500 mm in length, i.e. with five and 15 cells 100 mm in length, respectively

Figures 8.6b revel that for longer boards the finger-joints will, most of the time, resist more than the board material. For very short boards, however, this relation stops holding (see Fig. 8.6a), and might even reverse, depending on the specific distributions for $f_{t,0,\mathrm{cell}}$ and $f_{t,0,\mathrm{fj}}$. For the case studied here, this means that a very small effect should be observed in the bending strength of the simulated GLT beams, as the timber material is, statistically speaking, more decisive for the initiation of global failure of the beam.

Results

Figures 8.7a–f present the results for the parametric analysis of the length of boards. It can be seen that, as theorized above, the effect is very small, only distinguishable for shorter boards of 500 mm and, to a minor extent, also for lengths of 1000 mm. The differences are restricted mostly to the upper quantiles ($\gtrapprox$ 50 %-quantiles), while the lower tails behave very similarly.

It should be mentioned that the simulations do not consider the effect of cutting off weak regions (e.g. large knots) for shorter boards, and assumes a random shortening of boards instead. Nevertheless, as shown in Section 6.4.3, this approach delivers clearly higher strengths for shorter boards and lower strengths for longer boards. A more realistic simulation, where the cutting off of evidently weak regions is considered, would presumably show a larger effect than the presented simulation.

The size effect for the different configurations is analyzed in the following. Table 8.6 presents the computed size effect exponents, ξ, obtained for the GLT simulated with different board lengths. Beams made from shorter boards seem to

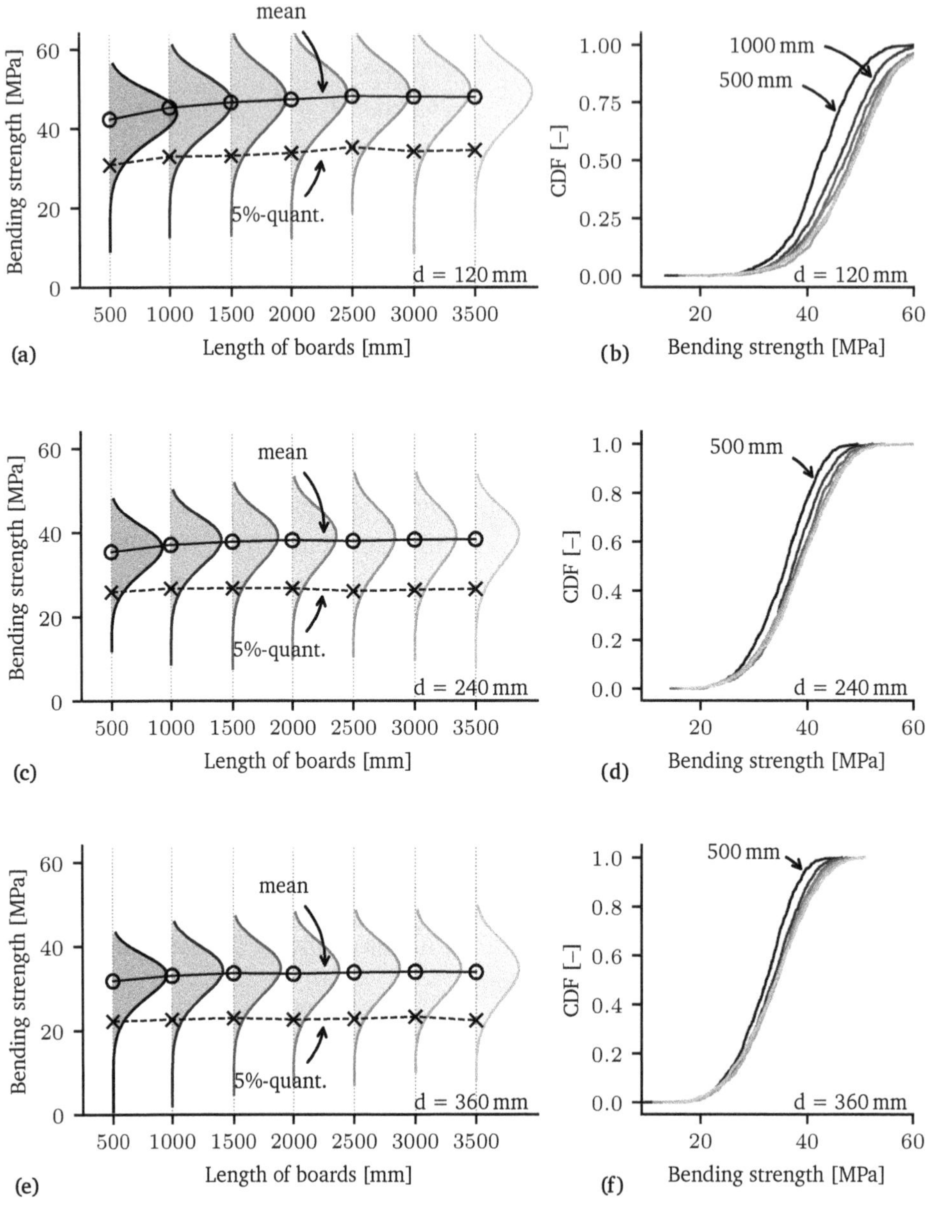

Figure 8.7. Glulam bending strength obtained for simulations with different board lengths and beam cross-sectional depth: (a,b) d = 120 mm, (c,d) d = 240 mm, (e,f) d = 360 mm

Table 8.6. Size effect computed for the simulated beams with different board lengths

	Size effect exponent, ξ [–]						
Board length [mm]	500	1000	1500	2000	2500	3000	3500
mean	0.26	0.29	0.30	0.32	0.32	0.31	0.32
5%-quant.	0.30	0.35	0.34	0.37	0.39	0.35	0.40

present a somewhat larger size effect as compared to using longer boards for the characteristic (5 %-quantile) values. On the mean level the computed size effect factors range between 0.35 and 0.38, and can be considered rather high for GLT beams. This stems from the rather low quality of the studied boards, as was also shown in Section 6.4.3, where the size effect for individual boards was numerically investigated.

Naturally, the results depend on the specific distributions of $f_{t,\mathrm{fj}}$ and $f_{t,0,\mathrm{cell}}$, but the general behavior should remain unchanged. This is, for shorter boards, normally having higher $f_{t,0}$, values, finger-joints become more relevant for the failure mechanism, whilst the opposite happens for longer boards.

This analysis indicates that shorter (stronger) boards account for the behavior of dataset B, where the finger-joints presented relatively low values compared to the tensile strength of the boards. These boards had a rather short length of about 600 mm, where most of the defects are cut out. The problem of *low strength* finger-joints can then be considered as a multidimensional problem, where an interaction of board length, classification criteria and bonding quality of the finger-joints defines the overall situation.

8.5.3 Effect of tensile strength of finger-joints

It has been shown that for GLT beams of oak, the failure mainly originates at the location of finger-joints, which is owed to the relative low strength of finger-joints as compared to the oak boards. This was experimentally observed in dataset B, where short, strong boards were used. Hence, it is of great interest to study how changes in the strength of finger-joints influence the global failure of the GLT beams when short boards ($\approx$ 600 mm) are used.

For the following analysis, the properties for $f_{t,0,\mathrm{cell}}$ and the distribution of board lengths are taken from dataset B. The distribution for $f_{t,\mathrm{fj}}$ is varied by modifying the generalized parameters "location" and "scale", producing a shifting and a change in spread, respectively. The original (base) distribution for the finger-joints of the outer LS13 laminations is described by a Lognormal distribution with loc = 24.7 MPa, scale = 15.2 and shape = 0.56. The location parameter is modified by adding or subtracting values ranging from −10 MPa to 25 MPa, in steps of 5 MPa; similarly, the scale parameter is modified by adding or subtracting values ranging from −10 to 10 in steps of 5.

An illustration of the modified distributions can be seen in Fig. 8.8a,b, where the base distribution of $f_{t,0}$ is presented, too. It can be seen from Fig. 8.8a that the chosen range of modified location parameters covers the entire $f_{t,0}$ distribution, which should translate in rather different behaviors for each configuration. In a similar way, Fig. 8.8b shows how the modified scale parameters span over the region of the $f_{t,0}$, distribution. As with the other analyses, a total of 1000 simulations are performed per configuration, and three different depths are analyzed: 6 and 9 and 15 laminations, corresponding to 120 mm, 200 mm and 300 mm, respectively.

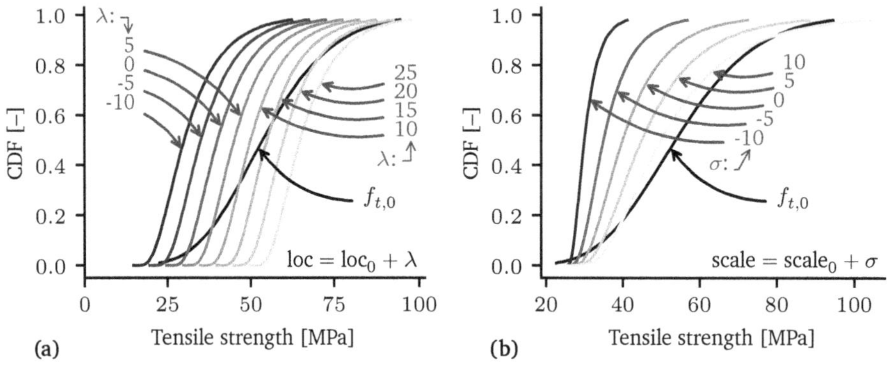

Figure 8.8. Investigated modified tensile strength, $f_{t,\text{fj}}$, distributions. (a) modification of the location parameter, (b) modification of the scale parameter

Results for the modified "location" parameter

Figures 8.9a–f present the results for the investigated "location" parameters for the different analyzed cross-sections. The mean and 5 %-quantile values are shown in Table 8.7. It can be seen that as the location parameter in increased (larger λ), the bending strength monotonically increases. However, the rate of increase decreases for larger λ values. This is expected, since, as the $f_{t,\text{fj}}$ distribution is shifted towards higher values, the wood material (boards) begins to gain more relevance—from the perspective of the failure mechanism. Eventually the point is reached where higher $f_{t,\text{fj}}$ values have no effect, as the failure mechanism is entirely controlled by the wood material.

Table 8.8 presents the percentage difference compared to the base case ($\lambda = 0$) for both, mean and 5 %-quantile levels. It is apparent that for larger depths the

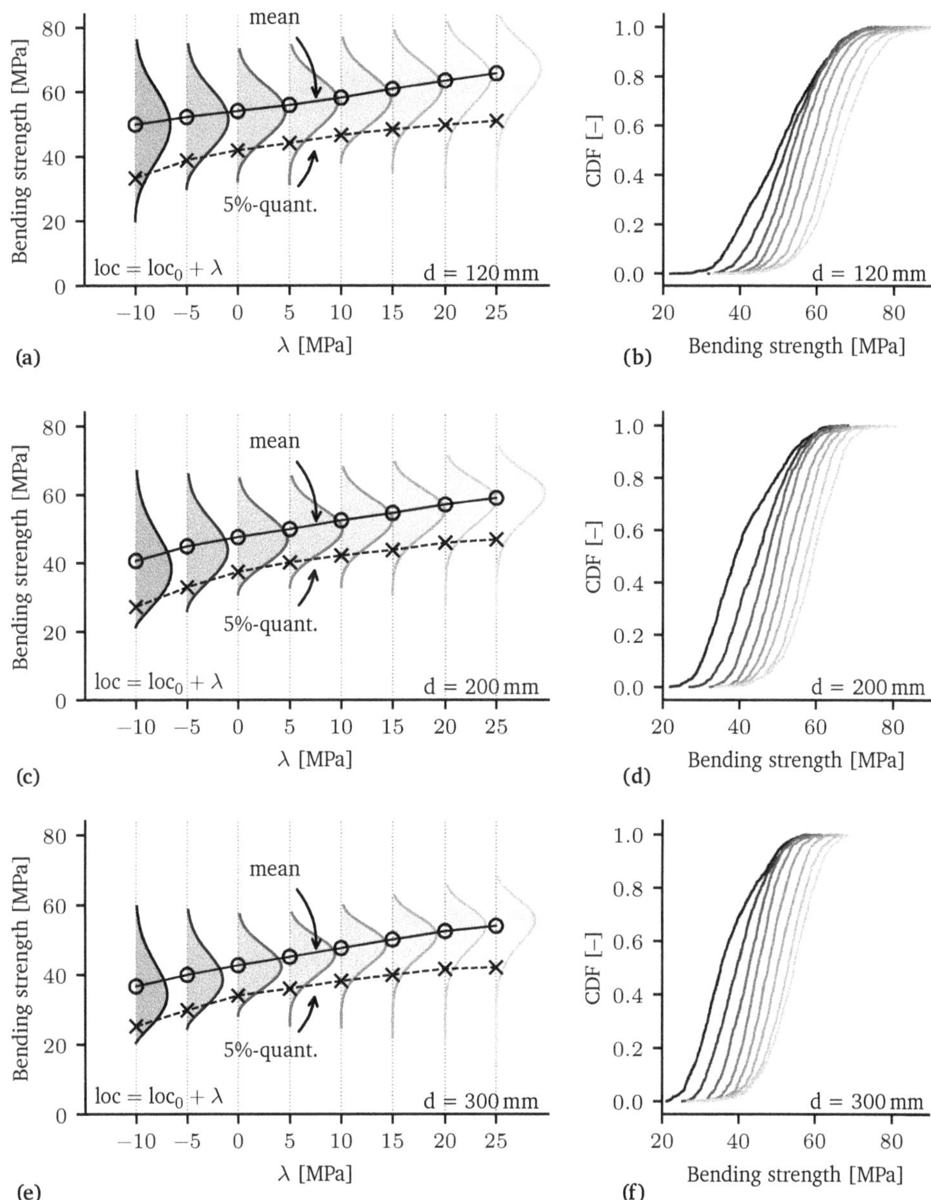

Figure 8.9. Simulated bending of GLT beams of different depths dependent on location parameter of the lognormal distribution describing the tensile strength of finger-joints. (a,b) d = 120 mm; (c,d) d = 200 mm; (e,f) d = 300 mm

Table 8.7. Mean and 5 %-quantile values for $f_{m,g}$ of different beam depths, resulting from the modification of the "location" parameter of the $f_{t,fj}$ distribution

Depth [mm]		Bending strength, $f_{m,g}$ [MPa]							
	λ [MPa]	-10	-5	0	5	10	15	20	25
120	mean	49.9	52.2	54.1	55.9	58.2	61.0	63.6	65.9
	5%-quant.	33.3	38.9	41.9	44.1	46.6	48.4	49.8	51.1
200	mean	40.6	44.9	47.5	49.9	52.4	54.6	57.1	59.0
	5%-quant.	27.3	33.0	37.4	40.2	42.2	43.8	45.9	46.9
300	mean	36.7	40.0	42.7	45.1	47.6	50.0	52.4	53.9
	5%-quant.	25.3	29.9	34.0	36.0	38.2	39.9	41.6	42.1

Table 8.8. Difference between the $f_{m,g}$ for different beam depths, obtained with the modified $f_{t,fj}$ distributions and the base case ($\lambda = 0$)

Depth [mm]		Difference relative to base case $\lambda = 0$ [%]							
	λ [MPa]	-10	-5	0	5	10	15	20	25
120	mean	−7.7	−3.4	0.0	3.4	7.7	12.8	17.6	21.8
	5%-quant.	−20.5	−7.3	0.0	5.3	11.2	15.5	18.7	21.8
200	mean	−14.5	−5.6	0.0	4.9	10.3	14.8	20.1	24.1
	5%-quant.	−27.1	−11.9	0.0	7.3	12.8	17.0	22.7	25.3
300	mean	−14.1	−6.4	0.0	5.7	11.5	17.2	22.8	26.3
	5%-quant.	−25.8	−12.2	0.0	5.8	12.2	17.3	22.3	23.6

effect of shifting the $f_{t,fj}$ distribution is stronger. This can be explained by the simple fact that, for longer beams the number of finger-joints increases, making it more likely for the weakest region to be determined by a finger-joint. In contrast, for smaller beams, fewer finger-joints are present, which increases the probability of having a weak region in the wood material. In the latter case, the improved $f_{t,fj}$ has a lesser impact as in the former case.

This information allows to assess the hypothetical improvement of the $f_{t,fj}$ distribution. For example, for an increase of 58 % in the mean value of $f_{t,fj}$—corresponding to $\lambda = 10$ MPa—, an increase of about 12 % can be expected for the 5 %-quantile value of the bending strength. This percentage depends, of course, on the specific distributions for $f_{t,0}$ and $f_{t,fj}$.

Finally, the size effect was computed for each configuration, based on the three analyzed beam depths. The results are presented in Table 8.9. It can be observed that the size effect decreases as the $f_{t,fj}$ distribution is shifted towards higher strength values. This is related to the percentage differences shown in Table 8.8. Since the same improvement in finger-joints produces a higher relative increase in $f_{m,g}$ for larger beams, the difference of strength between depths is reduced, thus resulting in the observed smaller size effect.

Table 8.9. Size effect exponent dependent on the location parameter of the lognormal of finger-joint distribution

	Size effect exponent, ξ [–]							
λ [MPa]	-10	-5	0	5	10	15	20	25
mean	0.33	0.29	0.26	0.24	0.22	0.22	0.21	0.22
5%-quant.	0.29	0.28	0.23	0.23	0.22	0.21	0.20	0.22

Table 8.10. Mean and 5 %-quantile values for $f_{m,g}$ of different beam depths, resulting from the modification of the "scale" parameter of the $f_{t,\text{fj}}$ distribution

		Bending strength, $f_{m,g}$ [MPa]				
	σ [–]	-10	-5	0	5	10
120	mean	49.9	52.3	54.1	56.2	58.2
	5%-quant.	36.0	39.9	41.8	43.6	44.2
200	mean	41.2	44.8	47.5	49.7	51.6
	5%-quant.	29.8	34.2	37.2	39.1	40.6
300	mean	37.2	40.4	42.8	44.8	46.6
	5%-quant.	27.1	31.4	34.0	35.5	36.6

Table 8.11. Difference between the $f_{m,g}$ for different beam depths, obtained with the modified $f_{t,\text{fj}}$ distributions and the base case ($\sigma = 0$)

		Difference relative to base case [%]				
	σ [–]	-10	-5	0	5	10
120	mean	−7.7	−3.3	0.0	4.0	7.7
	5%-quant.	−13.9	−4.7	0.0	4.3	5.6
200	mean	−13.3	−5.7	0.0	4.6	8.5
	5%-quant.	−20.0	−8.2	0.0	5.0	9.2
300	mean	−13.2	−5.7	0.0	4.6	8.8
	5%-quant.	−20.2	−7.7	0.0	4.5	7.8

Results for the modified "scale" parameter

Figures 8.10a–f and Talbe 8.10 present the results for $f_{m,g}$ concerning the aspect of "scale" parameter variation. For larger spreads, i.e. larger standard deviations, higher bending strength values are obtained. However, the effect decreases as the $f_{t,\text{fj}}$ distribution widens ($\sigma > 0$) and acquires similar characteristics as $f_{t,0}$ (see Fig. 8.8b). For higher values of the parameter σ, it is expected that mostly the wood material controls the failure mechanism.

The effect relative to the base configuration ($\sigma = 0$) is presented in Table 8.11 for both, mean and 5 %-quantile levels. Similarly as for the case of the location parameter variation, the modification of σ has a slightly larger effect for larger cross-sections. The explanation to this is very similar to the explanation for the

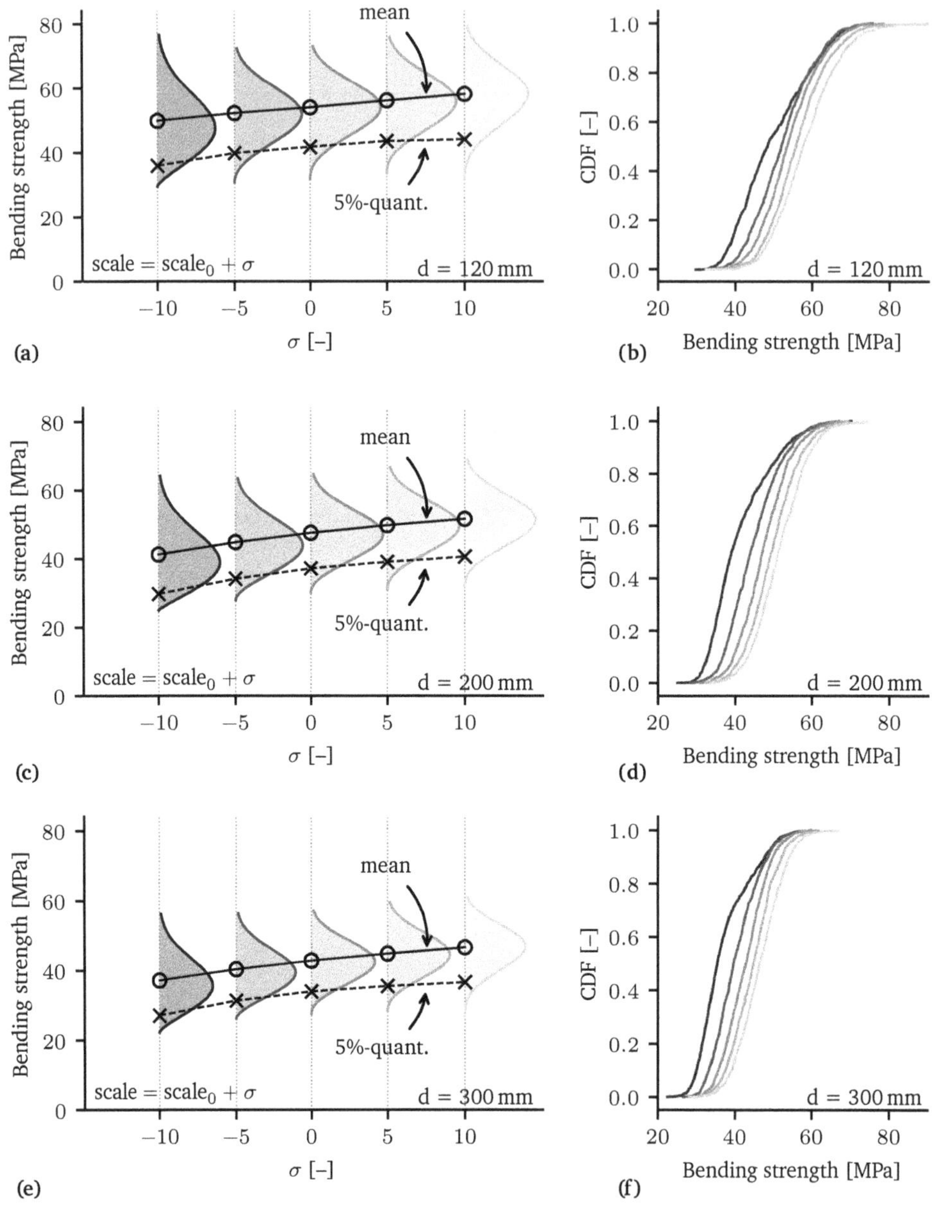

Figure 8.10. Simulated bending of GLT beams of different depths dependent on scale parameter of the lognormal distribution describing the tensile strength of finger-joints. (a,b) d = 120 mm; (c,d) d = 200 mm; (e,f) d = 300 mm

Table 8.12. Size effect computed for the simulated beams with different scale parameters for the distribution of finger-joints

	Size effect exponent, ξ [–]				
σ [–]	-10	-5	0	5	10
mean	0.31	0.28	0.26	0.25	0.24
5%-quant.	0.30	0.26	0.23	0.22	0.21

location parameter, namely the larger number of finger-joints present in longer beams.

The computed size effect exponents are shown in Table 8.12 for both, mean and 5 %-quantile levels. It can be seen that for larger spreads, the size effect decreases. This seems counterintuitive at first sight, however, this is owed to the specific analyzed case, where a larger spread of $f_{t,0,\text{fj}}$ resembles the distribution for $f_{t,0}$. This effectively translates into less variation of tensile strength throughout the simulated beam, as all the strength values might be regarded as following the same distribution instead of two different distributions.

8.6 Discussion

This final chapter presented some practical applications of the glulam strength model. After the initial calibration, the model was able to produce a good agreement with test results. The advantages of applying such a model are multiple, and discussed in the following.

The production of GLT beams and the declaration of its respective mechanical properties is based on the experience and results accumulated over many decades. This experience has lead to the development of a set of rules given in EN 14080 (2013), which ensure a reliable production of GLT beams, where the quality and safety of the product stand in the front line. These rules paved the way for the current ubiquity of softwood GLT, being the most used glued engineered element in timber constructions.

For the specific case of hardwood GLT, however, the accumulated experience is much smaller and clear rules are lacking for the definition of the relevant mechanical properties. In absence of clear rules—as in the case of softwoods—, the only alternative to reduce the uncertainty of the mechanical properties is normally considered to be the realization of experimental campaigns. Although experiments are normally regraded as the last word for such cases, the use of a strength model helps in the interpretation of the results, especially for the cases where a small number of specimens are tested.

The ultimate objective of this model is not to only to assist in the analysis of the results obtained from experimental campaigns, but, more ambitiously, reducing the number of required specimens to be tested—or even replacing them—, thus reducing their associate costs. Such approach would in essence be equivalent to the current approach according to EN 14080 (2013), however, the nature of the finite element model enables a higher degree of flexibility, allowing to make calibrations for specific species or predict characteristic values for specific loading conditions. It should be mentioned that the developed model can not only be used under bending conditions, but simulations under tension loading are straightforward to implement, similarly as done by Frese et al. (2017).

Furthermore, the strength model serves not only the prediction of bending strength based on empirically obtained parameters for material distributions, but also allows to assess the effects of possible improvements in the production chain, e.g. better finger-joints or better boards. This can be considered being a useful tool e.g. in the decision process whether or not to invest in a new finger-jointing machine—associated with a specific improvement in $f_{t,\mathrm{fj}}$—is justified, based on the specific criteria of an individual producer.

The most time-consuming part of applying the presented strength model is the determination of the correct parameters to describe the variation of the material properties within board. Ideally, this could be simplified by the specification of parameters describing the variation of MOE within board in a tabulated form. This could be done for specific strength grades or combination of grades, and for boards with different cross-sectional dimensions.

For the variation of strength along board, the presented method, based on survival analysis, can be used to estimate $f_{t,0,\mathrm{cell}}$. Preferably, multiple strength values per board should be obtained in order to better represent the real distribution of $f_{t,0,\mathrm{cell}}$. However, it was shown that the shape of the upper tail of $f_{t,0,\mathrm{cell}}$ has a rather small influence on the computed bending strengths. This means that suitable parameters might be derived based on one tensile test per board, perhaps in combination with an assumed maximum possible strength value, which could be fed to the parametric Beta model.

Conclusions and outlook

9.1 Conclusions

This thesis was initiated to address the general lack of knowledge regarding the variation of mechanical properties in hardwoods. This was addressed here for the particular case of European oak (*Q. robur*, *Q. petraea*). The mechanical properties along board are in general influenced by a multitude of variables, spanning from the microstructure of the wood cells to the larger scale defects. The interaction of the different variables is complex and in some cases the variables are itself unknown. Regression analyses do help to understand the influence of some of these variables and are very useful in the context of classification. However, these variables can be ignored and a pure aleatoric process can be assumed. In doing so, the focus shifts to the process itself, its nature, providing a new perspective of its characteristics. In the context of stochastic simulations, the latter approach shows some advantages, as only the needed process is simulated, eliminating the need to generate auxiliary variables like knots or density. The methodology applied to investigate the variation of properties along boards might, at times, have given the impression of being an exercise in pure statistics. However, the practical relevance of the introduced methods was hopefully made evident in the last chapter, where the methods were used in the developed finite element (FE) glulam strength model.

Although the final output of this work is the presented glulam strength model, it is evident that the study of the variation of properties within board constitutes the core of this thesis. The improvements of the developed FE strength model and similar models from the literature are, generally speaking, rather small. In fact, the main principles have not changed much since the first models presented in the 1980's, except for additional considerations as fracture mechanics and some adjustments in the discretization. Certainly the implementation of each model differs substantially, but the core remains unchanged. In contrast, the methodology adopted for the analysis of the variation of mechanical properties along board contains new ideas with a general character, thus making them applicable to different hardwood—and maybe softwood—species, too.

The conclusions deriving from the study of the variation of mechanical properties within board can be summarized as follows:

- The intra-board distribution of modulus of elasticity (MOE) presents a marked skew to the left, best represented by a Log-Gamma distribution. The observed skew points at a relatively large number of cells with high MOE values and a decreasing amount of cells with comparatively low MOE values. This behavior is also observed when large knots are removed, evidencing the rather large degree of variation that can be attributed to other factors such as the fiber deviation.
- The autoregressive analysis performed on the stationary MOE data strongly suggests that a first order autoregressive model [AR(1)] describes the data in a satisfactory manner.
- The computed lag-1 considering all cells and only clear wood segments were 0.44 and 0.58, respectively, highlighting the influence of large knots in the intra-board variation of MOE.
- The developed autoregressive model differs from most of the existing stochastic regression-based MOE variation approaches in that it is independent of knot indicators.
- The developed model is able to reproduce the observed statistical characteristics of the studied boards.
- The variation of tensile strength can be analyzed by means of survival analysis. The fitted models behave in agreement with the weakest link theory, i.e. simulating the length effect in a coherent manner.
- The cross-correlation between MOE and tensile strength models promotes lower tensile strength values in regions of low MOE and higher strength

in cells with higher relative MOE values, which is in agreement with the expected behavior.

- The simulated size-effect for oak boards is rather high compared to typical values for softwoods, showing an exponent between 0.30 and 0.33 for the mean level, and between 0.23 and 0.31 for the 5 %-quantile level depending on the used model. This seems to be a characteristic of oak boards, deriving from the rather large variation observed.

The following main conclusions can be drawn from the results obtained with the developed FE glulam strength model:

- The consideration of fracture mechanics improves the capacity of the glulam strength model to reproduce the observed behavior of experimental results.
- The calibrated fracture energy values of 11 N/mm and 15 N/mm for finger-joints and board material, respectively, delivered satisfactory results for both GLT datasets.
- The use of a structured file format to solve the complete set of parameters needed for a single GLT configuration and cryptographic hash functions to ensure reproducibility, reduces possible human errors related to changes in parameters.
- The model can be used to assess the impact of possible improvements to the quality of finger-joints or wood material on the bending strength of GLT beams.

9.2 Outlook

At the end of any research work, almost inevitably, some questions will remain unanswered. Furthermore, the insights gained during the methodological analysis and solving of one problem will, in many cases, uncover additional unknowns or motivate alternative approaches and applications. Although great effort is done in pursuing some of these ideas as they appear—producing new insights and new ideas—,time constraints or the prevention of an excessive widening of the scope will leave many open questions. Chapters need to be closed at some point. Hence, some ideas must be left for future research. In the following, possible research topics related to this thesis are discussed.

The method chosen to measure the localized MOEs is very time consuming. A different approach, which would also yield considerably higher resolutions, is the use of optical systems to measure strains by means of fiber glass cables. This method would allow for a faster and more reliable gathering of data, easily applicable to different datasets. The higher resolution translates into a more detailed autoregressive analysis, which can lead to an improvement of the MOE model.

The data analyzed here and the simulations derived from the fitted models indicate a significantly stronger size effect as compared to softwoods. However, the rather small size of the dataset prohibits a decisive claim with the needed degree of certainty. Thus, the size effect needs to be studied in a larger dataset of oak boards.

The largely known differences existing between different hardwood species makes it highly unlikely for the variation of mechanical properties along board of different hardwoods to be governed by the same parameters as obtained here. Considering the change in availability of natural resources, the fostering of glulam made of different hardwoods should be an objective for the timber industry. Therefore, the methods presented here, or similar, should be applied to boards of different hardwood species, with the objective of slowly building a database with parameters describing each species.

The glulam strength model can be used to assist the development and fitting of a design equation for glulam beams made of hardwood species, analogous to EN 14080 (2013). For this, a mixture of experimental results and simulations based on possible configurations of material properties can be made.

The glulam strength model can be modified and extended to consider different loading situations and failure mechanisms. For example, it is trivial to analyze the behavior of GLT under tension loading by simply changing the boundary conditions and extending the fracture behavior to the entire cross-section. Additionally, the variation of properties related to shear or tension perpendicular to grain could be considered, too. This would require modifications on the user-defined material subroutine implemented for Abaqus. There is, however, enough work done in the literature for the consideration of the interaction of multiple stress components, e.g. by Lukacevic et al. (2017) and Sandhaas et al. (2020).

Finally, given the large number of simulations required, the strength model could be reimplemented with a different software without licence restrictions (e.g. open-source). This would allow for a much larger parallelization of the computations, reducing the total time needed to study a specific configuration. Owed to the rather simple geometry, this should be possible to implement with existing tools in a moderate time frame.

References

Abaqus (2017). *ABAQUS/Standard User's Manual, Version 2017*. Providence, RI, USA: Dassault Systèmes.

Agestam, E., M. Karlsson, and U. Nilsson (2006). "Mixed Forests as a Part of Sustainable Forestry in Southern Sweden." In: *Journal of Sustainable Forestry* 21.2-3, pp. 101–117. DOI: 10.1300/J091v21n02_07.

Aicher, S. (1992). "Fracture energies and size effect law for spruce and oak in Mode I." In: *Workshop on determination of fracture properties of wood especially in Mode II and mixed Mode I and II — RILEM TC 133 Meeting*. Bordeaux, France.

Aicher, S., Z. Christian, and G. Dill-Langer (2014). "Hardwood glulams — Emerging timber products of superior mechanical properties." In: *Proc. World Conf. Timber Engineering, CD-ROM PAP7-11*. Quebec City, Canada. DOI: 10.13140/2.1.5170.1120.

Aicher, S., M. Hirsch, and Z. Christian (2016). "Hybrid cross-laminated timber plates with beech wood cross-layers." In: *Construction and Building Materials* 124, pp. 1007–1018. DOI: 10.1016/j.conbuildmat.2016.08.051.

Aicher, S. and G. Stapf (2014). "Glulam from European White Oak: Finger Joint Influence on Bending Size Effect." In: *Materials and Joints in Timber Structures*. Ed. by S. Aicher, H.-W. Reinhardt, and H. Garrecht. Dordrecht: Springer Netherlands, pp. 641–656. ISBN: 978-940-077-8-1-1-5. DOI: 10.1007/978-94-007-7811-5_58.

Akaike, H. (1974). "A new look at the statistical model identification." In: *IEEE Transactions on Automatic Control* 19.6, pp. 716–723. ISSN: 0018-9286. DOI: 10.1109/TAC.1974.1100705.

ASTM D245 (2019). *Standard Practice for Establishing Structural Grades and Related Allowable Properties for Visually Graded Lumber*. West Conshohocken, PA, United States: ASTM International. DOI: 10.1520/d0245-06r19.

ASTM D6570-18a (2018). *Standard Practice for Assigning Allowable Properties for Mechanically Graded Lumber*. West Conshohocken, PA, United States: ASTM International.

Barrett, J. D. and A. R. Fewell (1990). “Size factors for the bending and tension strength of structural timber.” In: *International council for building research studies and documentation. Working Comission W18A – Timber Structures. Meeting 23*. Lisbon, Portugal.

Bechtel, F. K. (1985). “Beam stiffness as a function of pointwise E, with application to machine stress rating.” In: *Proc. Int. Smp. Forest Prod. Res.* CSIR. Pretoria, South Africa.

Blank, L., G. Fink, R. Jockwer, and A. Frangi (2017). “Quasi-brittle fracture and size effect of glued laminated timber beams.” In: *European Journal of Wood and Wood Products*, pp. 667–681. ISSN: 1436-736X. DOI: 10.1007/s00107-017-1156-0.

Blaß, H. J. and M. Frese (2006). *Biegefestigkeit von Brettschichtholz-Hybridträgern mit Randlamellen aus Buchenholz und Kernlamellen aus Nadelholz*. German. Tech. rep. Karlsruher Institut für Technologie (KIT). 55 pp. DOI: 10.5445/KSP/1000005148.

Blaß, H. J., M. Frese, P. Glos, P. Linsenmann, and J. Denzler (2005). *Biegefestigkeit von Brettschichtholz aus Buche*. German. Tech. rep. Karlsruher Institut für Technologie (KIT). 141 pp. DOI: 10.5445/KSP/1000001371.

Brockwell, P. J. and R. A. Davis (2002). *Introduction to Time Series and Forecasting*. Springer International Publishing. DOI: 10.1007/978-3-319-29854-2.

Burger, N. and P. Glos (1996). “Einfluß der Holzabmessungen auf die Zugfestigkeit von Bauschnittholz.” In: *Holz als Roh- und Werkstoff* 54.5, pp. 333–340. DOI: 10.1007/s001070050197.

Castillo, E., A. S. Hadi, N. Balakrishnan, and J. M. Sarabia (2005). *Extreme Value and Related Models with Applications in Engineering and Science*. John Wiley & Sons, Inc. ISBN: 0-471-67172-X.

CEC (2007). *Green Paper from the Comission to the Council, the European Parliament, the European Economic and Social Committee and the Committee of the Regions. Adapting to Climate Change in Europe – Options for EU Action*. Brussels, Belgium: Commission of the European Communities. URL: https://eur-lex.europa.eu/legal-content/en/ALL/?uri=CELEX:52007DC0354.

Chatterjee, S. and D. L. McLeish (1986). “Fitting linear regression models to censored data by least squares and maximum likelihood methods.” In: *Communications in Statistics — Theory and Methods* 15.11, pp. 3227–3243. DOI: 10.1080/03610928608829305.

Cheaib, A., V. Badeau, J. Boe, I. Chuine, C. Delire, et al. (2012). "Climate change impacts on tree ranges: model intercomparison facilitates understanding and quantification of uncertainty." In: *Ecology Letters* 15.6, pp. 533–544. DOI: 10.1111/j.1461-0248.2012.01764.x.

Chen, S. S. and R. A. Gopinath (2000). "Gaussianization." In: *Advances in Neural Information Processing Systems 13, Papers from Neural Information Processing Systems (NIPS) 2000, Denver, CO, USA*, pp. 423–429. URL: http://papers.nips.cc/paper/1856-gaussianization.

Colling, F. (1990a). "Beigefestigkeit von Brettschichtholzträgern in Abhängigkeit von den festigkeitsrelevanten Einflußgrößen." In: *European Journal of Wood and Wood Products* 48.9, pp. 321–326.

Colling, F. and M. Scherberger (1987). "Die Streuung des Elastizitätsmoduls in Brettlängsrichtung." In: *Holz als Roh und Werkstoff* 45, pp. 95–99.

Colling, F. (1990b). "Tragfähigkeit von Brettschichtholz in Abhängigkeit von den festigkeitsrelevanten Einflussgrößen." PhD thesis. Germany: Universität Fridericiana zu Karlsruhe.

Davidson-Pilon, C., J. Kalderstam, P. Zivich, B. Kuhn, A. Fiore-Gartland, et al. (June 2019). *CamDavidsonPilon/lifelines: v0.21.3*. DOI: 10.5281/zenodo.3240536.

Delzon, S., M. Urli, J.-C. Samalens, J.-B. Lamy, H. Lischke, F. Sin, N. E. Zimmermann, and A. J. Porté (2013). "Field Evidence of Colonisation by Holm Oak, at the Northern Margin of Its Distribution Range, during the Anthropocene Period." In: *PLOS ONE* 8.11, pp. 1–8. DOI: 10.1371/journal.pone.0080443.

Deodatis, G. and R. C. Micaletti (2001). "Simulation of Highly Skewed Non-Gaussian Stochastic Processes." In: *Journal of Engineering Mechanics* 127.12, pp. 1284–1295. DOI: 10.1061/(ASCE)0733-9399(2001)127:12(1284).

Dill-Langer, G., R. C. Hidalgo, F. Kun, Y. Moreno, S. Aicher, and H. J. Herrmann (July 2003). "Size dependency of tension strength in natural fiber composites." In: *Physica A: Statistical Mechanics and its Applications* 325.3-4, pp. 547–560. DOI: 10.1016/s0378-4371(03)00141-9.

DIN 4074-5 (2008). *Strength grading of wood – Part 5: Sawn hard wood*. Berlin, Germany: German Institute for Standardization.

EN 384 (2016). *Statistical evaluation of test results MOR, MOE and density with structural sized spruce log specimens tested in axial compression at dry and wet (water saturated) state*. Brussels, Belgium: Comité Européen de Normalisation.

EN 408 (2012). *Timber structures – Structural timber and glued laminated timber – Determination of some physical and mechanical properties*. Brussels, Belgium: Comité Européen de Normalisation.

EEC (1988). *Council Directive 89/106/EEC of 21 December 1988 on the approximation of laws, regulations and administrative provisions of the Member States relating to construction products*. Brussels, Belgium. URL: `http://data.europa.eu/eli/dir/1989/106/oj`.

Ehlbeck, J. and F. Colling (1987). *Biegefestigkeit von Brettschichtholz in Abhängigkeit von Rohdichte, Elsatizitätsmodul, Ästigkeit und Keilzinkung der Lamellen, der Lage der Keilzinkung sowie von der Trägerhöhe*. Tech. Rep. Karlsruhe, Germany: Versuchsanstalt für Stahl, Holz und Steine, Universität Fridericiana Karlsruhe.

Ehlbeck, J., F. Colling, and R. Görlacher (1985). "Einfluß keilgezinkter Lamellen auf die Biegefestigkeit von Brettschichtholzträgern." In: *Holz als Roh- und Werkstoff* 43.8, pp. 333–337. ISSN: 1436-736X. DOI: `10.1007/BF02607817`.

Ehlbeck, J., F. Colling, and R. Görlacher (1984). *Einfluß keilgezinkter Lamellen auf die Biegefestigkeit von Brettschichtholzträgern*. Tech. rep. Karlsruhe, Germany: Versuchsanstalt für Stahl, Holz und Steine, Universität Fridericiana Karlsruhe.

Ehrhart, T. (June 2020). "European beech (Fagus sylvatica L.) glued laminated timber: lamination strength grading, production and mechanical properties." In: *Holz als Roh- und Werkstoff*. DOI: `10.1007/s00107-020-01545-6`.

EN 1194 (1999). *Timber structures — Glued laminated timber — Strength classes and determination of characteristic values*. European Committee for Standardization.

EN 14080 (2013). *Timber structures – Glued laminated timber and glued solid timber – Requirements*. Brussels, Belgium: European Committee for Standardization.

EN 14081-1 (2016). *Timber structures – Strength graded structural timber with rectangular cross section – Part 1: General requirements*. Brussels, Belgium: European Committee for Standardization.

EN 14358 (2016). *Timber structures – Calculation and verification of characteristic values*. Brussels, Belgium: Comité Européen de Normalisation.

EN 1995-1-1 (2010). *Eurocode 5: Design of timber structures – Part 1-1: General – Common rules and rules for buildings*. Brussels, Belgium: European Committee for Standardization.

EN 975-1 (2009). *Sawn timber – Appearance grading of hardwoods – Part 1: Oak and beech*. Brussels, Belgium: European Committee for Standardization.

ETA-13/0642 (2013). *European Technical Approval, "VIGAM – Glued laminated timber of oak"*. Vienna, Austria: Österreichisches Institut für Bautechnik.

ETA-13/0646 (2013). *European Technical Approval, "SIEROLAM – Glued laminated timber of chestnut"*. Vienna, Austria: Österreichisches Institut für Bautechnik (OIB).

EU (2014). *Regulation (EU) No 305/2011 of the European Parliament and of the Council of 9 March 2011 laying down harmonised conditions for the marketing of construction products and repealing Council Directive 89/106/EEC (Text with*

EEA relevance). (Consolidated text). Brussels, Belgium. URL: http://data.europa.eu/eli/reg/2011/305/2014-06-16.

EU Hardwoods (2017). *European hardwoods for the building sector*. Final report. URL: https://forestvalue.org/wp-content/uploads/2018/07/wwnet_jc4_final_reporting_eu_hardwoods.pdf.

Faydi, Y., L. Brancheriau, G. Pot, and R. Collet (2017). "Prediction of Oak Wood Mechanical Properties Based on the Statistical Exploitation of Vibrational Response." In: *BioResources* 12.3. ISSN: 1930-2126.

Faye, C., G. Legrand, D. Reuling, and J. .-. Lanvin (2017). "Experimental investigations on the mechanical behaviour of glued laminated beams made of oak." In: *International Network on Timber Engineering Research (INTER) – Meeting 50*. Kyoto, Japan: Timber Scientific Publishing, Karlsruhe, Germany, pp. 193–206.

Felton, A., M. Lindbladh, J. Brunet, and Ö. Fritz (2010). "Replacing coniferous monocultures with mixed-species production stands: An assessment of the potential benefits for forest biodiversity in northern Europe." In: *Forest Ecology and Management* 260.6, pp. 939–947. ISSN: 0378-1127. DOI: 10.1016/j.foreco.2010.06.011.

Fink, G., M. Deublein, and J. Kohler (2011). "Assessment of different knot-indicators to predict strength and stiffness properties of timber boards." In: *Proceedings of the 44th Meeting, International Council for Research and Innovation in Building and Construction, Working Commission W18 – Timber Structures,* Paper No. 44–1.

Fink, G. (2014). "Influence of varying material properties on the load-bearing capacity of glued laminated timber." PhD thesis. Zürich, Switzerland: ETH Zürich. DOI: 10.3929/ethz-a-010108864.

Foley, C. (2003). "Modeling the effects of knots in Structural Timber." PhD thesis. Lund University.

Foschi, R. O. (1987). "A procedure for the determination of localized modulus of elasticity." In: *Holz Roh- Werkstoff* 45, pp. 257–260.

Foschi, R. O. and J. D. Barrett (1980). "Glued-laminated beam strength: A model." In: *ASCE J. Struct. Div.* 106, pp. 1735–1754.

Frese, M. (2006a). "Brettschichtholz aus Buche – Ein leistungsfähiger Baustoff." German. In: *Ingenieurholzbau, Karlsruher Tage 2006. Forschung für die Praxis; 5. und 6. Oktober 2006; Tagungsband. Universität Karlsruhe, Lehrstuhl für Ingenieurholzbau und Baukonstruktionen*. Bruderverlag, Karlsruhe, pp. 28–39.

Frese, M. (2006b). "Die Biegefestigkeit von Brettschichtholz aus Buche. Experimentelle und numerische Untersuchungen zum Laminierungseffekt." German. PhD thesis. 152 pp. ISBN: 3-86644-043-X. DOI: 10.5445/KSP/1000004599.

Frese, M., S. Egner, and H. J. Blaß (2017). "Reliability of large glulam members – Part 2: Data for the assessment of partial safety factors for the tensile strength." In: *International Network on Timber Engineering Research (INTER) – Meeting 50*. Kyoto, Japan: Timber Scientific Publishing, Karlsruhe, pp. 453–470.

Frühwald, K. and G. Schickhofer (2005). "Strength Grading of Hardwoods." In: *14th International Symposium on Nondestructive Testing of Wood*. Vol. 14. Eberswalde, Germany, pp. 199–210.

Glos, P. and B. Lederer (2000). *Sortierung von Buchen-und Eichenschnittholz nach der Tragfähigkeit und Bestimmung der zugehörigen Festigkeits-und Steifigkeitskennwerte*. Inst. für Holzforschung.

Görlacher, R. (1990). "Klassifizierung von Brettschichtholzlamellen durch Messung von Longitudinalschwingungen." PhD thesis. Universität Fridericiana zu Karlsruhe.

Gričar, J., M. de Luis, P. Hafner, and T. Levanič (2013). "Anatomical characteristics and hydrologic signals in tree-rings of oaks (Quercus robur L.)" In: *Trees* 27.6, pp. 1669–1680. DOI: 10.1007/s00468-013-0914-9.

Grigoriu, M. (1998). "Simulation of Stationary Non-Gaussian Translation Processes." In: *Journal of Engineering Mechanics* 124.2, pp. 121–126. DOI: 10.1061/(ASCE)0733-9399(1998)124:2(121).

Hernandez, R., D. A. Bender, B. A. Richburg, and K. S. Kline (1992). "Probabilistic Modeling of Glued-Laminated Timber Beams." In: *Wood and Fiber Sciences* 24 (3), pp. 294–306.

Hu, M., A. Olsson, M. Johansson, and J. Oscarsson (2018). "Modelling local bending stiffness based on fibre orientation in sawn timber." In: *European Journal of Wood and Wood Products* 76.6, pp. 1605–1621. ISSN: 1436-736X. DOI: 10.1007/s00107-018-1348-2.

Isaksson, T. (1999). "Modelling the Variability of Bending Strength in Structural Timber – Length and Load Configuration Effects." PhD thesis. Lund Institute of Technology – Division of Structural Engineering.

Kandler, G. and J. Füssl (2017). "A probabilistic approach for the linear behaviour of glued laminated timber." In: *Engineering Structures* 148, pp. 673–685. ISSN: 0141-0296. DOI: 10.1016/j.engstruct.2017.07.017.

Kandler, G., J. Füssl, and J. Eberhardsteiner (2015a). "Stochastic finite element approaches for wood-based products: theoretical framework and review of methods." In: *Wood Science and Technology* 49.5, pp. 1055–1097. ISSN: 1432-5225. DOI: 10.1007/s00226-015-0737-5.

Kandler, G., J. Füssl, E. Serrano, and J. Eberhardsteiner (2015b). "Effective stiffness prediction of GLT beams based on stiffness distributions of individual lamellas."

In: *Wood Science and Technology* 49.6, pp. 1101–1121. ISSN: 1432-5225. DOI: 10.1007/s00226-015-0745-5.

Kandler, G., M. Lukacevic, and J. Füssl (2016). "An algorithm for the geometric reconstruction of knots within timber boards based on fibre angle measurements." In: *Construction and Building Materials* 124, pp. 945–960. ISSN: 0950-0618. DOI: 10.1016/j.conbuildmat.2016.08.001.

Kandler, G., M. Lukacevic, S. Wolff, and J. Füssl (2018). "Stochastic engineering framework for timber structural elements." In: *Beton- und Stahlbetonbau* 113.S2, pp. 96–102. DOI: 10.1002/best.201800055.

Kaplan, E. L. and P. Meier (1958). "Nonparametric Estimation from Incomplete Observations." In: *Journal of the American Statistical Association* 53.282, pp. 457–481. ISSN: 01621459. URL: http://www.jstor.org/stable/2281868.

Kline, D., F. Woeste, and B. Bendtsen (1986). "Stochastic model for modulus of elasticity of lumber." In: *Wood and Fiber Science* 18(2), pp. 228–238.

Klöck, W. (2005). "Statistical analysis of the shear strength of glued laminated timber based on full-size flexure tests." In: *Otto-Graf-Journal*. Vol. 16. Stuttgart, Germany: Otto-Graf-Institute (FMPA), University of Stuttgart.

Kollmann, F. F. P. and W. A. Côté (1968). *Principles of Wood Science and Technology: I Solid Wood*. 1st ed. Springer-Verlag Berlin Heidelberg. ISBN: 978-3-642-87930-2,978-3-642-87928-9.

Lam, F., R. O. Foschi, J. D. Barrett, and Q. Y. He (1993). "Modified algorithm to determine localized modulus of elasticity of lumber." In: *Wood Science and Technology* 27.2, pp. 81–94. ISSN: 1432-5225. DOI: 10.1007/BF00206227.

Lam, F. and E. Varoglu (1990). "Effect of length on the tensile strength of lumber." In: *Forest products Journal* 40.5, pp. 37–42.

Lam, F. and E. Varoglu (1991a). "Variation of tensile strength along the length of lumber – Part 1: Experimental." In: *Wood Science and Technology* 25.5, pp. 351–359. DOI: 10.1007/BF00226174.

Lam, F. and E. Varoglu (1991b). "Variation of tensile strength along the length of lumber – Part 2: Model development and verification." In: *Wood Science and Technology* 25.6, pp. 449–458. DOI: 10.1007/BF00225237.

Lam, F., J. D. Barrett, and S. Nakajima (2005). "Influence of knot area ratio on the bending strength of Canadian Douglas fir timber used in Japanese post and beam housing." In: *Journal of Wood Science* 51.1, pp. 18–25. DOI: 10.1007/s10086-003-0619-6.

Lam, F., Y.-T. Wang, and J. D. Barrett (1994). "Simulation of Correlated Nonstationary Lumber Properties." In: *Journal of Materials in Civil Engineering* 6.1, pp. 34–53. DOI: 10.1061/(ASCE)0899-1561(1994)6:1(34).

Lanvin, J.-D. and D. Reuling (2012). "Mechanical evaluation of French oak (Sessile and Pendunculate) for structural use." In: *Revue Forestière Française* LXIV.2. (in French), pp. 151–165. DOI: 10.4267/2042/47474.

Larsen, H. J. (1980). *Strength of glued laminated beams. Part 5*. Report No. 8004. Aalborg, Denmark: Institute of Building Technology and Structural Engineering, Aalborg University.

Lexer, M. J., K. Hönninger, H. Scheifinger, C. Matulla, N. Groll, H. Kromp-Kolb, K. Schadauer, F. Starlinger, and M. Englisch (2002). "The sensitivity of Austrian forests to scenarios of climatic change: a large-scale risk assessment based on a modified gap model and forest inventory data." In: *Forest Ecology and Management* 162.1. National and Regional Climate Change Impact Assessments in the Forestry Sector, pp. 53–72. ISSN: 0378-1127. DOI: 10.1016/S0378-1127(02)00050-6.

LFBW (2014). *Richtlinie Landesweiter Waldentwicklungstypen*. Ed. by G. Wicht-Lückge. Kernerplatz 10, 70182 Stuttgart: Landesbetrieb Forst Baden-Württemberg; Ministerium für Ländlichen Raum und Verbraucherschutz Baden-Württemberg. URL: www.mlr.baden-wuerttemberg.de.

Lindbladh, M., R. Bradshaw, and B. H. Holmqvist (2000). "Pattern and Process in South Swedish Forests during the Last 3000 Years, Sensed at Stand and Regional Scales." In: *Journal of Ecology* 88.1, pp. 113–128. ISSN: 00220477, 13652745. URL: http://www.jstor.org/stable/2648489.

Lindner, M., J. B. Fitzgerald, N. E. Zimmermann, C. Reyer, S. Delzon, et al. (2014). "Climate change and European forests: What do we know, what are the uncertainties, and what are the implications for forest management?" In: *Journal of Environmental Management* 146, pp. 69–83. ISSN: 0301-4797. DOI: 10.1016/j.jenvman.2014.07.030.

Lukacevic, M., J. Füssl, and J. Eberhardsteiner (2018). "Simulating Failure Mechanisms in Wooden Boards with Knots by Means of a Microstructure-Based Multisurface Failure Criterion." In: *CD-ROM Proceedings of the World Conference on Timber Engineering (WCTE 2018)*. Seoul, Republic of Korea: WCTE.

Lukacevic, M., G. Kandler, M. Hu, A. Olsson, and J. Füssl (2019). "A 3D model for knots and related fiber deviations in sawn timber for prediction of mechanical properties of boards." In: *Materials & Design* 166, p. 107617. ISSN: 0264-1275. DOI: 10.1016/j.matdes.2019.107617.

Lukacevic, M., W. Lederer, and J. Füssl (2017). "A microstructure-based multisurface failure criterion for the description of brittle and ductile failure mechanisms of clear-wood." In: *Engineering Fracture Mechanics* 176, pp. 83–99. ISSN: 0013-7944. DOI: 10.1016/j.engfracmech.2017.02.020.

Madsen, B. (1990). "Length effects in 38 mm spruce-pine-fir dimension lumber." In: *Canadian Journal of Civil Engineering* 17.2, pp. 226–237. DOI: 10.1139/l90-028.

Meier, E. S., H. Lischke, D. R. Schmatz, and N. E. Zimmermann (2011). "Climate, competition and connectivity affect future migration and ranges of European trees." In: *Global Ecology and Biogeography* 21.2, pp. 164–178. DOI: 10.1111/j.1466-8238.2011.00669.x.

Miller, R. G. (1976). "Least Squares Regression with Censored Data." In: *Biometrika* 63.3, pp. 449–464. ISSN: 00063444. DOI: 10.2307/2335722.

NF B 52-001-1 (2011). *Classement visuel pour l'emploi en structures des bois sciés français résineux et feuillus*. Association Française de Normalisation.

Odell, P. M., K. M. Anderson, and R. B. D'Agostino (1992). "Maximum Likelihood Estimation for Interval-Censored Data Using a Weibull- Based Accelerated Failure Time Model." In: *Biometrics* 48.3, pp. 951–959. ISSN: 0006341X, 15410420. DOI: 10.2307/2532360.

Olsson, A., J. Oscarsson, E. Serrano, B. Källsner, M. Johansson, and B. Enquist (2013). "Prediction of timber bending strength and in-member cross-sectional stiffness variation on the basis of local wood fibre orientation." In: *European Journal of Wood and Wood Products* 71.3, pp. 319–333. ISSN: 1436-736X. DOI: 10.1007/s00107-013-0684-5.

Olsson, A., G. Pot, J. Viguier, Y. Faydi, and J. Oscarsson (2018). "Performance of strength grading methods based on fibre orientation and axial resonance frequency applied to Norway spruce (Picea abies L.), Douglas fir (Pseudotsuga menziesii (Mirb.) Franco) and European oak (Quercus petraea (Matt.) Liebl./Quercus robur L.)" In: *Annals of Forest Science* 75.4, p. 102. ISSN: 1297-966X. DOI: 10.1007/s13595-018-0781-z.

Paviot, T. (2018). *pythonOCC, 3D CAD/CAE/PLM development framework for the Python programming language, PythonOCC – 3D CAD Python*. (accessed May 10, 2018). URL: http://www.pythonocc.org/.

Rao, R. V., D. P. Aebischer, and M. P. Denne (1997). "Latewood Density in Relation to Wood Fibre Diameter, Wall Thickness, and Fibre and Vessel Percentages in Quercus Robur L." In: *IAWA Journal* 18.2, pp. 127–138. DOI: 10.1163/22941932-90001474.

Reiterer, A., G. Sinn, and S. E. Stanzl-Tschegg (2002). "Fracture characteristics of different wood species under mode I loading perpendicular to the grain." In: *Materials Science and Engineering: A* 332.1, pp. 29–36. ISSN: 0921-5093. DOI: 10.1016/S0921-5093(01)01721-X.

Reiterer, A., S. E. Stanzl-Tschegg, and E. K. Tschegg (2000). "Mode I fracture and acoustic emission of softwood and hardwood." In: *Wood Science and Technology* 34.5, pp. 417–430. ISSN: 1432-5225. DOI: 10.1007/s002260000056.

Ross, R. J. (2010). *Wood handbook: wood as an engineering material*. U.S. Department of Agriculture, Forest Service, Forest Products Laboratory. DOI: 10.2737/fpl-gtr-190.

Rouger, F. and J. D. Barrett (1995). "Size effects in timber." In: *Informationsdienst Holz, STEP 3*, pp. 3/1–3/24.

Sandhaas, C., A. Khaloian, and J.-W. van de Kuilen (2020). "Numerical modelling of timber and timber joints: computational aspects." In: *Wood Science and Technology* 54, pp. 31–61. DOI: 10.1007/s00226-019-01142-8.

Seidl, R., M.-J. Schelhaas, and M. J. Lexer (2011). "Unraveling the drivers of intensifying forest disturbance regimes in Europe." In: *Global Change Biology* 17.9, pp. 2842–2852. DOI: 10.1111/j.1365-2486.2011.02452.x.

Serrano, E. and H. J. Larsen (1999). "Numerical Investigations of the Laminating Effect in Laminated Beams." In: *Journal of Structural Engineering* 125.7, pp. 740–745. DOI: 10.1061/(ASCE)0733-9445(1999)125:7(740).

Showalter, K., F. Woeste, and B. Bendtsen (1987). *Effect of Length on Tensile Strength in Structural Lumber*. Research Report FPL-RP-482. Madison, USA: U.S. Department of Agriculture, Forest Service, Forest Products Laboratory.

Song, J.-K., S.-Y. Kim, and S.-W. Oh (2007). "The Compressive Stress-strain Relationship of Timber." In: *Proceedings of the International Conference on Sustainable Building Asia*. Seoul, Korea, pp. 977–982.

Stapf, G. (Aug. 2010). "Energiedissipationsfähigkeit und Ermüdungsverhalten von geklebten Holzverbindungen." Diplomarbeit. Universität Kassel.

Stefanou, G. (2009). "The stochastic finite element method: Past, present and future." In: *Computer Methods in Applied Mechanics and Engineering* 198.9, pp. 1031–1051. ISSN: 0045-7825. DOI: 10.1016/j.cma.2008.11.007.

Stefanou, G. and M. Papadrakakis (Apr. 2007). "Assessment of spectral representation and Karhunen–Loève expansion methods for the simulation of Gaussian stochastic fields." In: *Computer Methods in Applied Mechanics and Engineering* 196.21-24, pp. 2465–2477. DOI: 10.1016/j.cma.2007.01.009.

Tapia, C. (2021). *Pybaqus: A Python library for the visualization and post-processing of Abaqus ASCII result files*. Version 0.2.1. DOI: 10.5281/zenodo.6373291.

Tapia, C. (2022). *Source code for: Stuttgart Stochastic Strength Glulam Model (S3GluM)*. Version V1. DOI: 10.18419/darus-1155.

Tapia, C. and S. Aicher (2020). *Replication Data for: Simulation of the localized modulus of elasticity of hardwood boards by means of an autoregressive model.*

Version V2. DOI: 10.18419/darus-863. URL: https://doi.org/10.18419/darus-863.
Tapia, C. and S. Aicher (2021). *Replication Data for: Survival analysis of tensile strength variation and simulated length-size effect along oak boards*. Version V2. DOI: 10.18419/darus-864.
Taylor, S. and D. Bender (1988). "Simulating Correlated Lumber Properties Using a Modified Multivariate Normal Approach." In: *Trans. ASAE* 31 (1), pp. 182–186. DOI: 10.13031/2013.30685.
Taylor, S. and D. Bender (1991). "Stochastic model for localized tensile strength and modulus of elasticity in lumber." In: *Wood and Fiber Science* 23(4), pp. 501–519.
Tong, Y. L. (1990). *The multivariate normal distribution*. English. 1st ed. New York, USA: Springer-Verlag. ISBN: 978-1-4613-9655-0.
Virtanen, P., R. Gommers, T. E. Oliphant, M. Haberland, T. Reddy, et al. (2020). "SciPy 1.0: Fundamental Algorithms for Scientific Computing in Python." In: *Nature Methods* 17, pp. 261–272. DOI: 10.1038/s41592-019-0686-2.
Weibull, W. (1951). "A Statistical Distribution Function of Wide Applicability." In: *Journal of Applied Mechanics* 18, pp. 293–297.
Wright, S., J. Dahlen, C. Montes, and T. L. Eberhardt (2019). "Quantifying knots by image analysis and modeling their effects on the mechanical properties of loblolly pine lumber." In: *European Journal of Wood and Wood Products*. ISSN: 1436-736X. DOI: 10.1007/s00107-019-01441-8.
Z-9.1-704 (2012). *Glulam made of oak. Holder of approval: Elaborados y Fabricados Gámiz. S.A., Spain*. Berlin, Germany: German Institute for Building Technology (DIBt).
Z-9.1-821 (2013). *Holz Schiller oak post and beam glulam. Holder of approval: Holz Schiller GmBH, Germany*. Berlin, Germany: German Institute for Building Technology (DIBt).

List of Symbols

Latin letters

a_i Diameter of knots, measured on surface of board perpendicular to grain

b Width of cross-section of board

d Depth of GLT beam

$f(\cdot)$ Probability density function

$f_{c,0}$ Compressive strength of boards

$f_{m,g}$ Bending strength of GLT beam

$f_{m,j}$ Flat-wise bending strength of finger-joints

$f_{t,0,\mathrm{cell}}$ Cell-wise tensile strength parallel to grain

$f_{t,0,\mathrm{glob}}$ Global tensile strength of board

$f_{t,0,\mathrm{sec}}$ Tensile strength obtained from secondary tensile tests

$f_{t,0}$ Tensile strength of board perpendicular to grain

$f_{t,0}$ Tensile strength of boards

t Thickness of cross-section of board

x_{05} 5 %-quantile level

$\bar{E}_{\mathrm{max},3}$ Average of three largest $E_{t,0,\mathrm{cell}}$ values in a board

$\widetilde{E}_{t,0,i,j}$ Normalized MOE of cell j of board i

$A_{i,k}$ List with tuples of k-lagged values per board

$C_{i,k}$ List with the concatenation of the $A_{i,k}$ lists of each board

E_{dyn} Dynamic MOE parallel to grain

$E_{m,0}$ Flat-wise bending stiffness of boards

$E_{m,g}$ Bending stiffness of GLT beam

$E_{t,0,\text{cell}}$ Cell-wise MOE parallel to grain direction

$E_{t,0,\text{glob}}$ Global MOE parallel to grain direction in board

$E_{t,0,i,j}$ Measured MOE of cell j of board i

$E_{t,0}$ MOE parallel to grain direction

$F(\cdot)$ Cumulative distribution function

$F_{\text{min}}(\cdot)$ Distribution of minimum value

G_f Fracture energy

L Length of window used to analyze knots along board

P Load

R Coefficient of correlation

Z Random variates corresponding to a standard normal distribution

$Z_{t,0,\text{cell}}$ Stationary data corresponding to MOE variation within board

$\text{KAR}_{\text{thres}}$ KAR threshold used for analysis of clear wood segments

ℓ_{board} Length of board

ℓ_{cell} Length of cell

ℓ_{CW} Length of clear wood segment

ℓ_{knots} Length used to measure knot variables

$\ell_{b,\text{mean}}$ Mean length of boards in GLT

$\ell_{E,\text{glob}}$ Length used to measure the global MOE in board

ℓ_g Length used to measure grain deviation

$\ell_{s,2,i}$ Testing length of the i-th secondary tensile tests of boards

ℓ_s Free length in tensile test

$\mathcal{L}(\cdot)$ Likelihood function

$\mathcal{N}$ Normal distribution

Greek letters

α_0 Intercept of the regression for scale parameter δ in Weibull and Beta regression models

α_1 Regression coefficient associated with $E_{t,0,\text{glob}}$ for parameter scale δ in Weibull and Beta regression models

β_0 Intercept of the regression for shape parameters ρ and a in Weibull and Beta regression models, respectively

β_1 Regression coefficient associated with $E_{t,0,\text{glob}}$ for parameter scale ρ and a in Weibull and Beta regression models, respectively

δ_i Binary variable indicating whether the i-th value is censored or not

$\gamma(\cdot)$ Autocovariance function

$\hat{\gamma}(\cdot)$ Sample autocovariance function

$\hat{\rho}(\cdot)$ Sample autocorrelation function

μ Mean value

ν Frequency

$\rho(\cdot)$ Autocorrelation function

ρ Density

ρ_0 Density corrected at 0 % MC

ρ_{12} Density corrected at 12 % MC

σ Standard deviation

σ_{tot} Total standard deviation of cell-wise MOE in a group of boards

$\sigma_{E,\text{glob}}$ Standard deviation of global MOE of boards

$\sigma_{E,\text{local}}$ Standard deviation of cell-wise MOE for one board

$\sigma_{t,0}$ Tensile stress parallel to grain

θ Parameters defining a given distribution function

ε_i	White noise component of AR process
φ_i	Model parameters for AR model
$\underline{\mu}$	Vector of mean values of a multinomial distribution
$\hat{\theta}$	Estimated parameters for a given distribution function
Δ	Used to denote a difference
$\Gamma(\cdot)$	Gamma function
$\mathbf{\Sigma}$	Covariance matrix
$\Phi(\cdot)$	Cumulative distribution function of the standard normal distribution
$\Phi^{-1}(\cdot)$	Percent point function of the standard normal distribution

Superscripts

cell	Variable computed from measured cell-wise results
mean	Mean value of measured variables
test	Variable measured directly from test results

Subscripts

0	Property parallel to grain
100	Value associated to a window of 100 mm
12	Property corrected to 12 % MC
150	Value associated to a window of 150 mm
c	Property under compressive conditions
j	Property associated to finger-joints
k	Characteristic value
t	Property under tensile conditions
cell	Property associated to one cell in board
glob	Global property, associated to entire board
max	Maximum value
min	Minimum value

Abbreviations

ACF	Autocorrelation function
AIC	Aikaike Information Criterion
AR	Autoregressive model
CDF	Cumulative distribution function
COV	Coefficient of correlation
CWAR	Clear wood area ratio
CWS	Clear wood segment
GLT	Glued laminated timber
IP	Indicating property
KAR	Knot area ratio
MC	Moisture content
MOE	Modulus of elasticity
PDF	Probability density function
SACF	Sample autocorrelation function
WL	Weakest link
WS	Weak segment

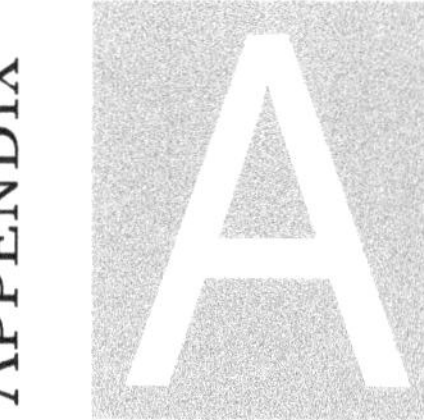

VARIATION OF MOE ALONG BOARDS

The entirety of the analyzed results regarding the variation of the modulus of elasticity parallel to the fiber direction, and clear wood ratios is presented here. The board No. 85 was not used in the analysis because a software-related malfunctioning of the testing machine brought it to an early (compressive) failure, before the global MOE could be measured. Therefore, no data for MOE or tensile strength is available. Nevertheless, the results for the MOE and CWAR variation along the board are presented in the following appendix (Fig. A.4c), since the author estimates that it adds relevant information to the understanding of the MOE variation along boards.

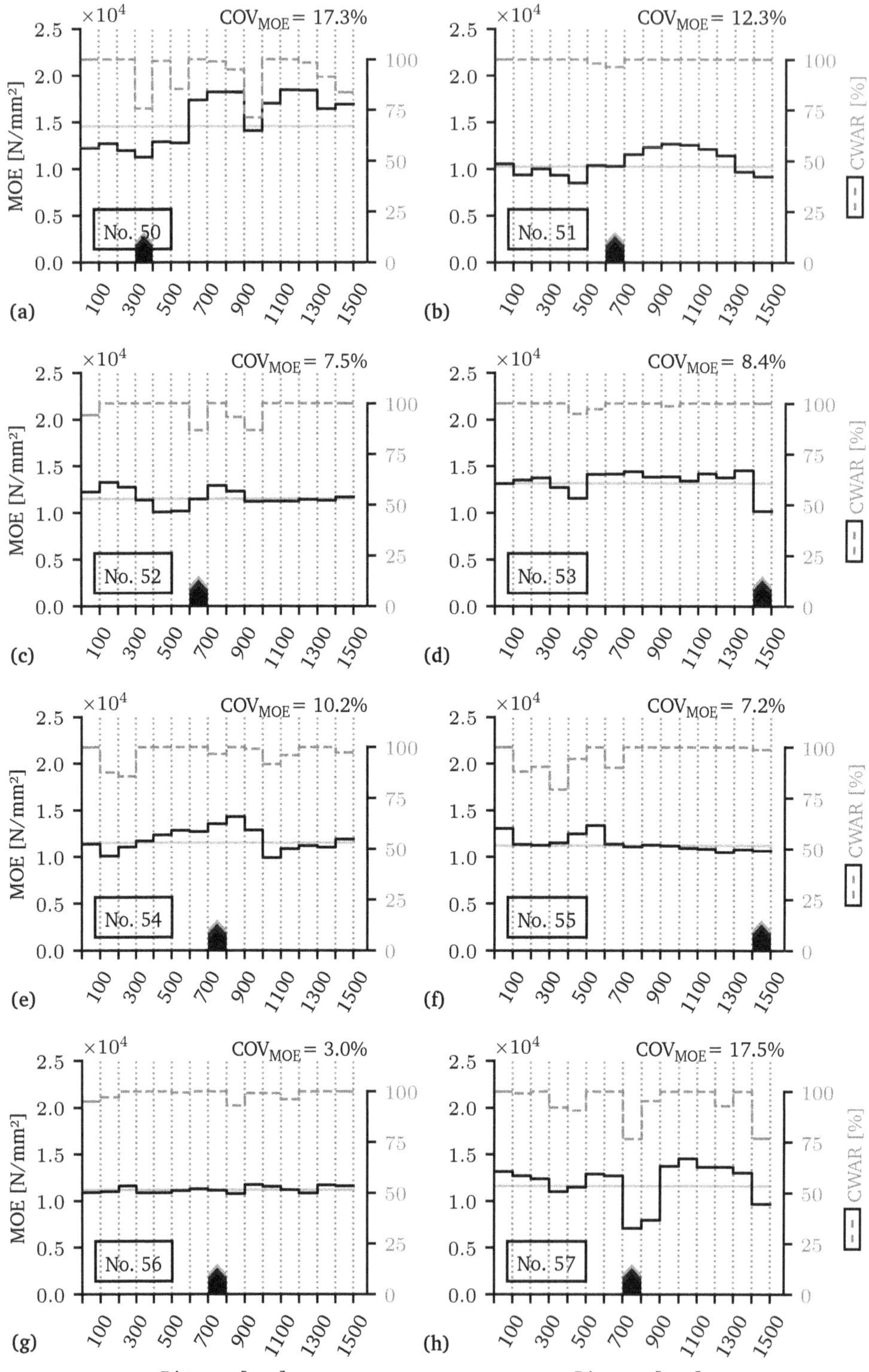

Figure A.1. MOE and CWAR variation for boards 50, 51, 52, 53, 54, 55 and 56. Horizontal line represents the measured global MOE. Black wedge marks the location of the global failure.

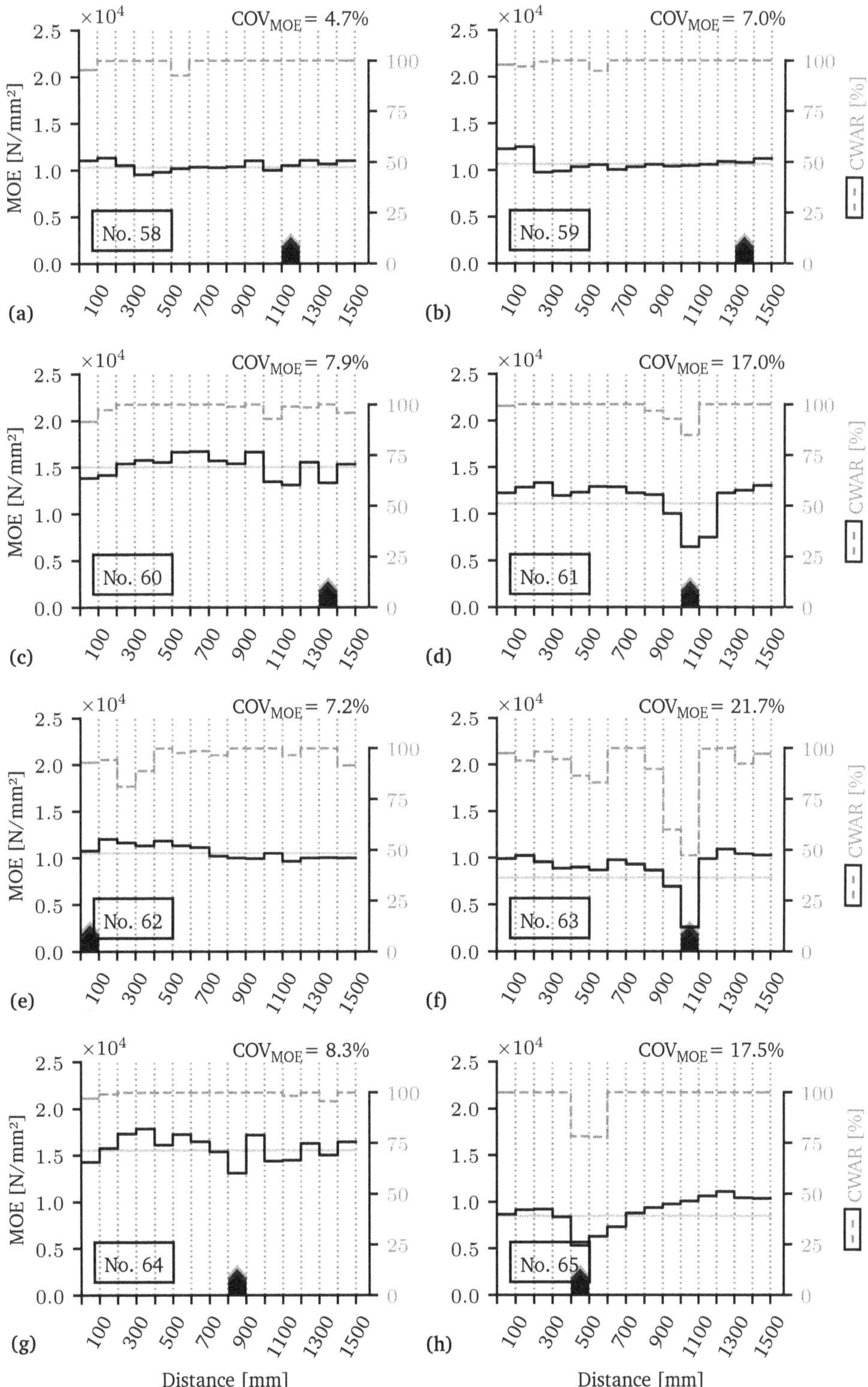

Figure A.2. MOE and CWAR variation for boards 58, 59, 60, 61, 62, 63, 64 and 65. Horizontal line represents the measured global MOE. Black wedge marks the location of the global failure.

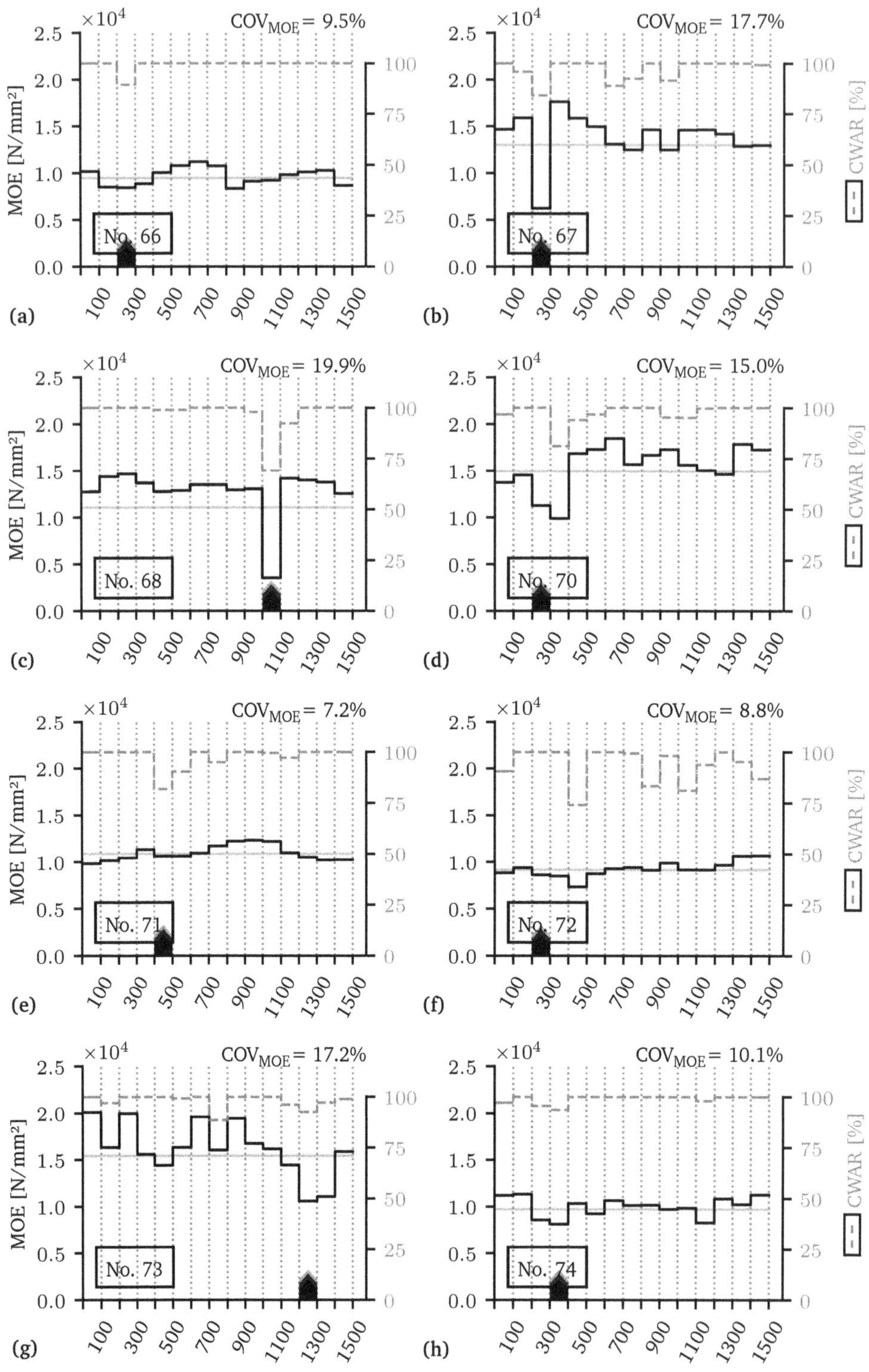

Figure A.3. MOE and CWAR variation for boards 66, 67, 68, 70, 71, 72, 73 and 74. Horizontal line represents the measured global MOE. Black wedge marks the location of the global failure.

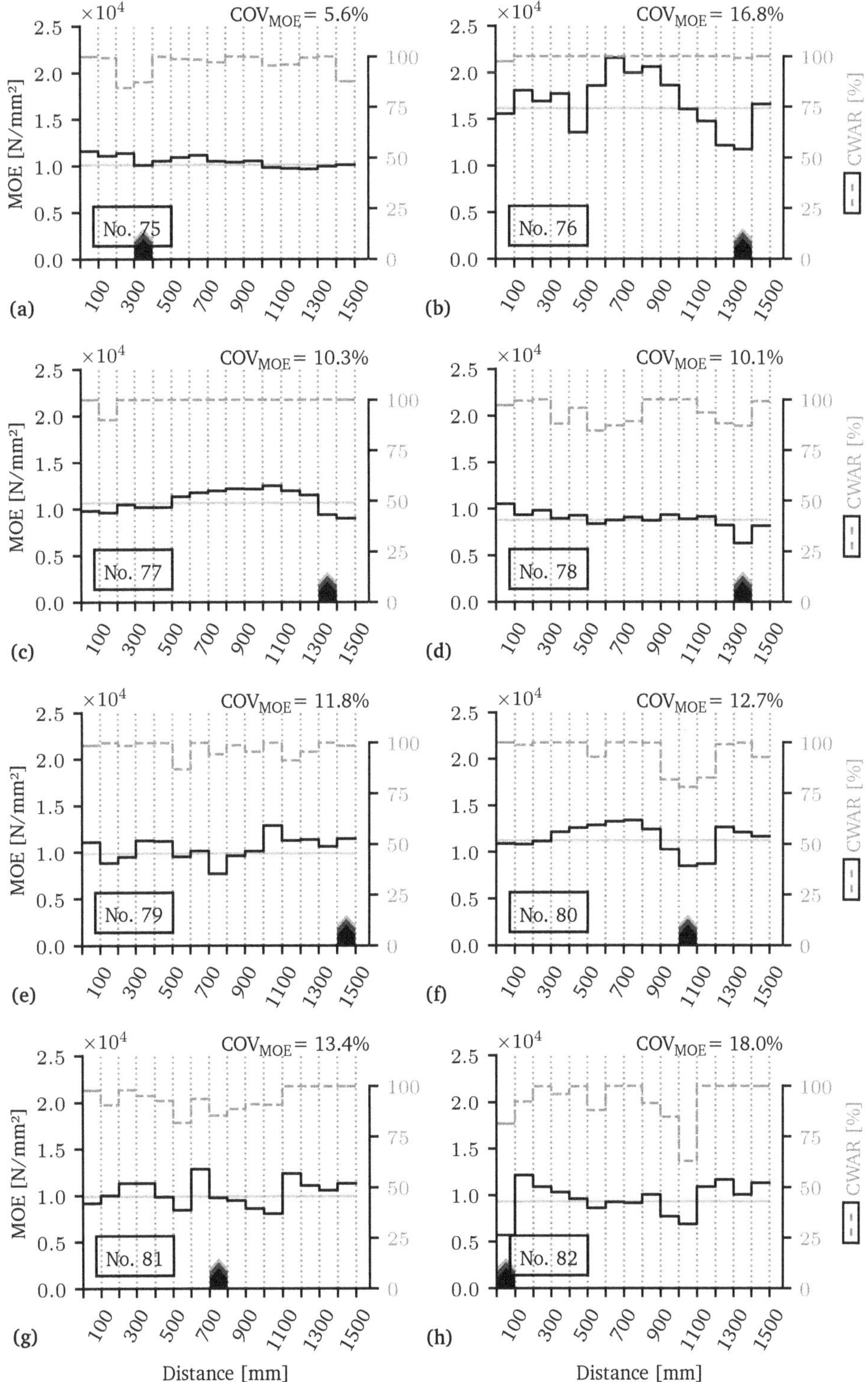

Figure A.4. MOE and CWAR variation for boards 75, 76, 77, 78, 79, 80, 81 and 82. Horizontal line represents the measured global MOE. Black wedge marks the location of the global failure.

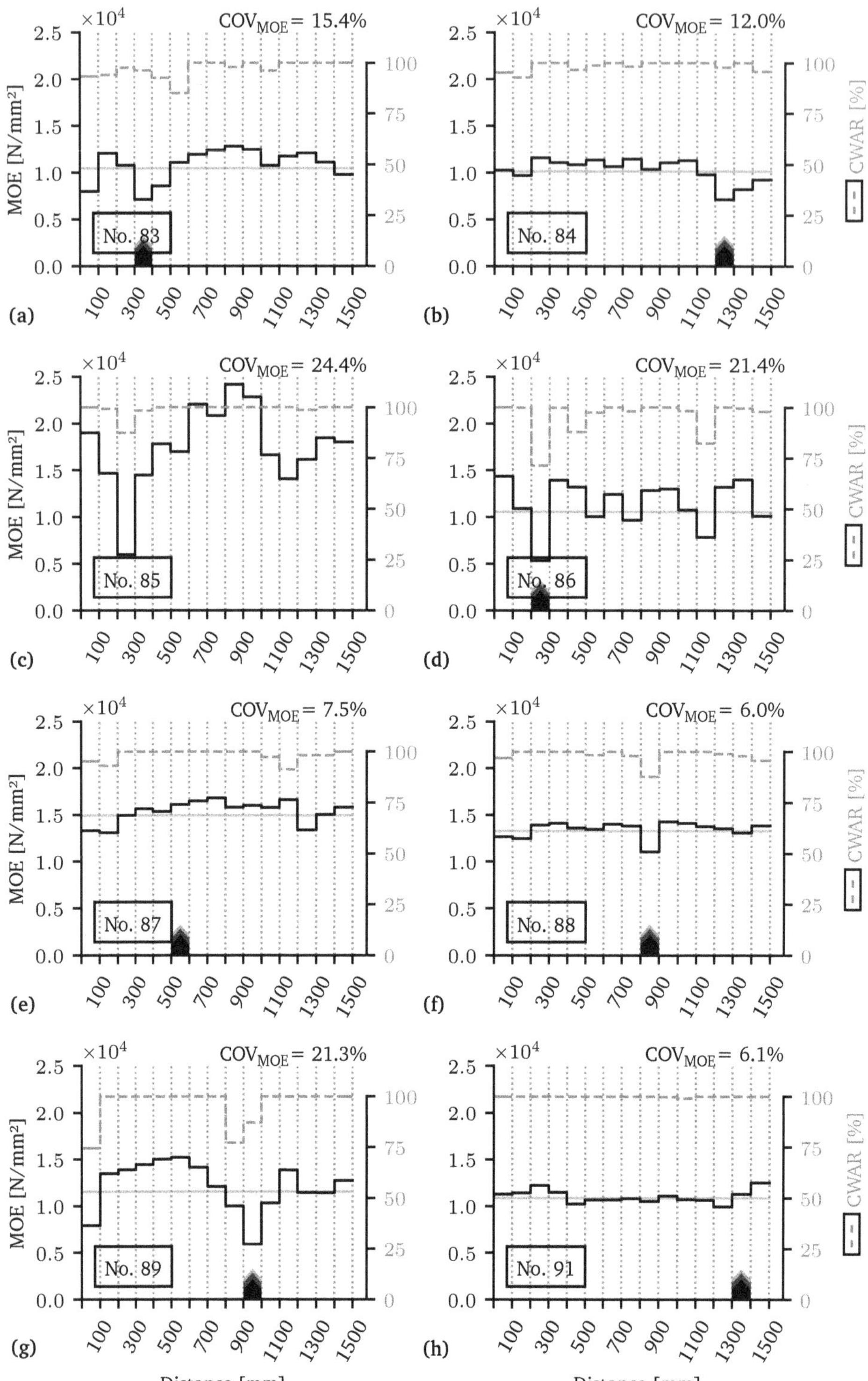

Figure A.5. MOE and CWAR variation for boards 83, 84, 85, 86, 87, 88, 89 and 91. Horizontal line represents the measured global MOE. Black wedge marks the location of the global failure.

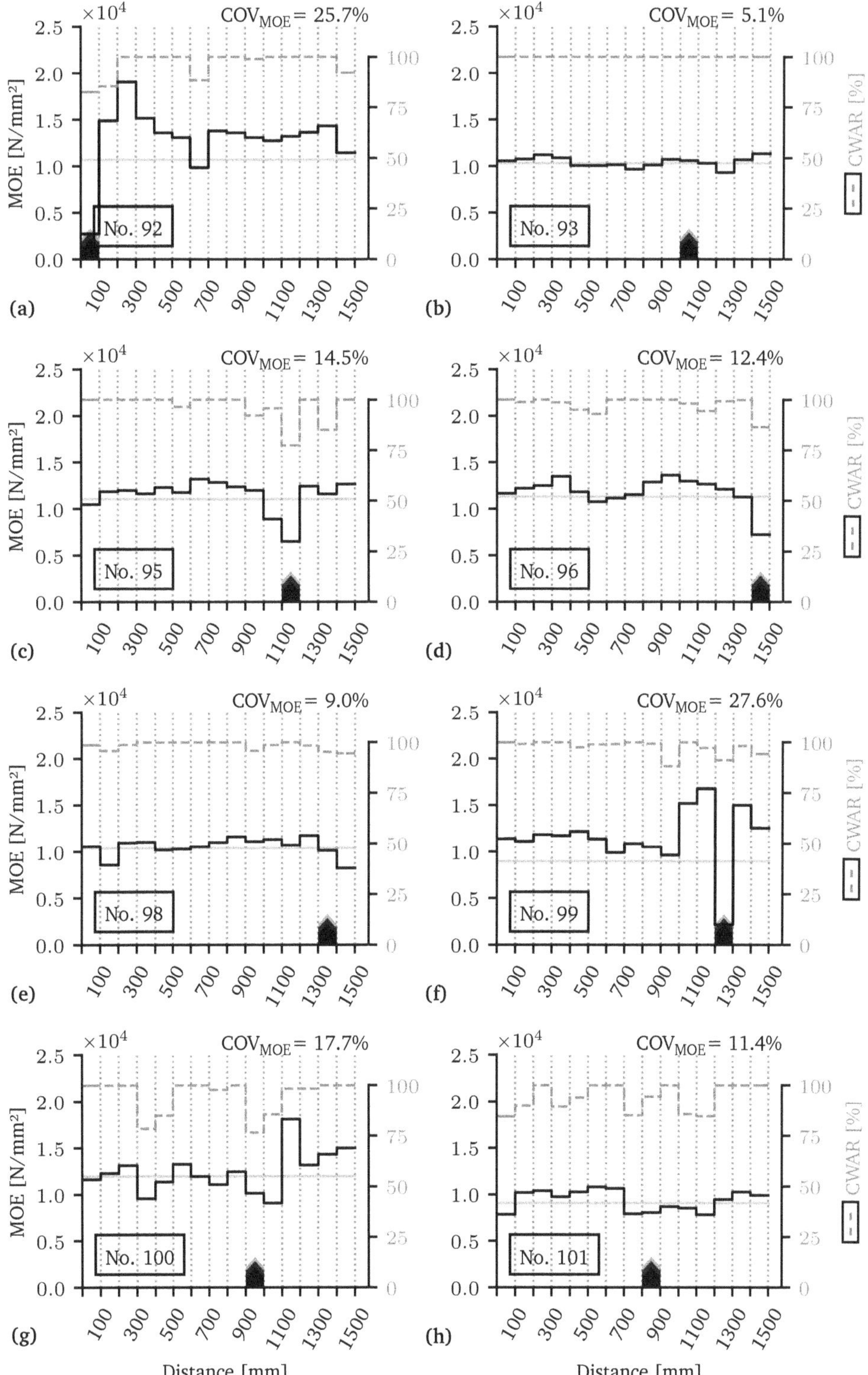

Figure A.6. MOE and CWAR variation for boards 92, 93, 95, 96, 98, 99, 100 and 101. Horizontal line represents the measured global MOE. Black wedge marks the location of the global failure.

Appendix

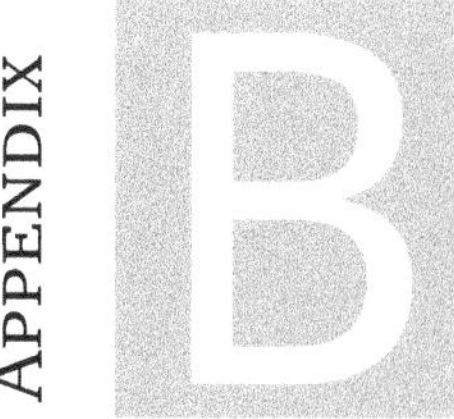

Analysis of MOE normalization method

The objective of this annex is to show the suitability of the MOE normalization method presented in Section 5.4.4. Specifically, it will be shown, by means of a simple numerical example, that the normalization method does not distort the shape of the original distribution. Thus, demonstrating that the observed skew is not an artifact of the chosen method.

The numerical example consists of a series of simple simulations, where the normalization method is used on distributions with different skews. In order to demonstrate this, Fig. B.1 (top) presents three Weibull distributions with three different shape parameters (1.5, 3.0 and 10.0). Each distribution is used to generate 1000 samples of 15 randomly generated values each (representing the studied boards). These *fake boards* are then normalized as done with the experimental data. Figure B.1 (bottom) shows the obtained normalized distributions for each of the above-lying original distributions, where it is evident that the normalization procedure does not affect the original skewness in any noticeable manner. Thus, it can be said that the chosen normalization method respects the underlying distribution of MOE within board and does not affect its general characteristics (i.e. the skewness is largely preserved). This holds as long as the support of the distribution is positive.

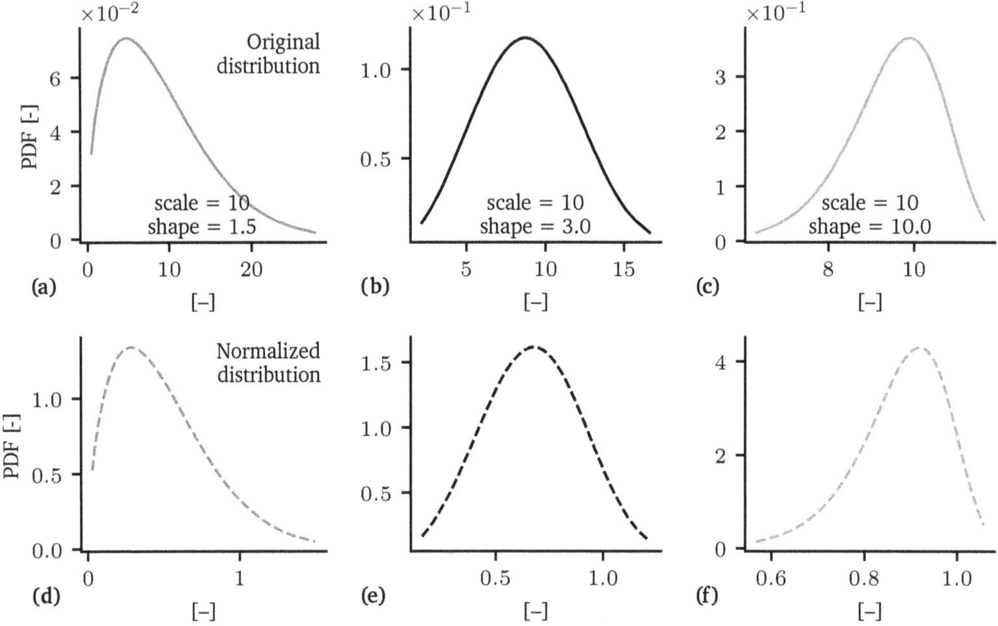

Figure B.1. Normalization method tested on Weibull distributions with three different shape parameters: (a) 1.5, (b) 3.0, and (c) 10.0; (d,e,f) distributions resulting from applying the normalization method to each original distribution

TESTING OF THE USER-DEFINED MATERIAL IN A ONE-ELEMENT PROBLEM

The used-defined material subroutine (UMAT) was tested with one element loaded in one direction by means of a set of given displacements. A two-dimensional, plain stress element of type CPS4R was used. The dimensions of the element were 10 × 10 mm. A tensile strength of $f_{t,0} = 40$ MPa was defined and the fracture energy was set as $G_f = 6$ N/mm. The material properties used are listed in Table C.1. The displacements were defined as a load-unload-load series, depicted in Fig. C.2a.

Figure C.2b shows the stress-strain result for the loading-unloading cycle. It can be seen that the element responds as expected, where after reaching $f_{t,0}$ softening starts. Once the load is reduced, the element unloads linearly exhibiting

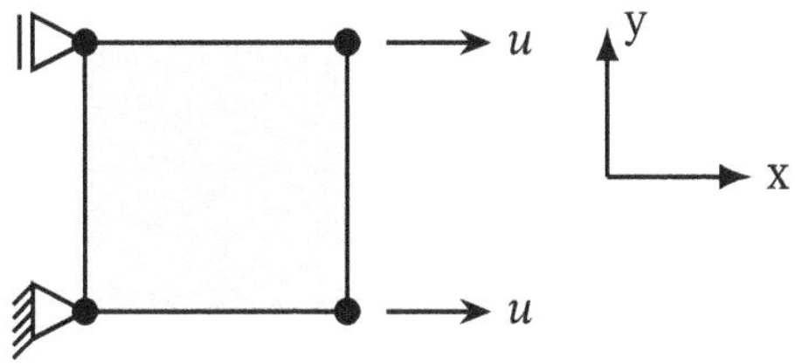

Figure C.1. Model of the tested configuration with one element

Table C.1. Material properties used for the one-element model

E_0 [MPa]	E_{90} [MPa]	ν_{xy} [–]	G_{xy} [MPa]	$f_{t,0}$ [MPa]	G_f [MPa]
12 500	350	0.02	650	40	4

the original elasticity. When the element is loaded again, it takes load until it reaches the previous load level prior to the unloading. Thus, it can be said that the implementation of the material with softening capabilities satisfies the requirements for the simulations.

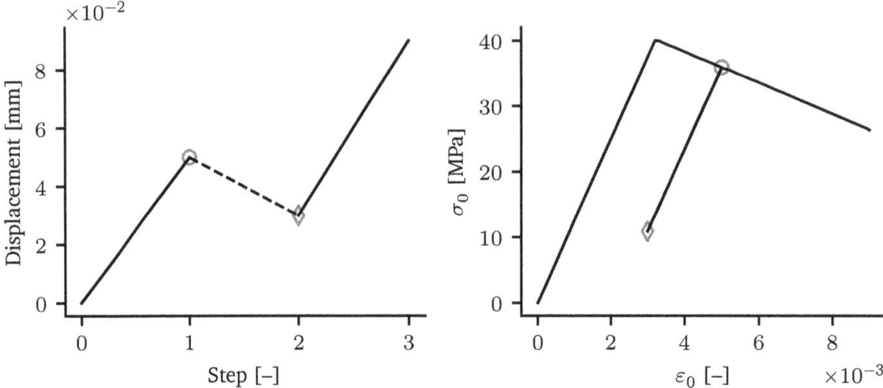

Figure C.2. Results for the one-element test with the UMAT subroutine. (a) applied displacement to the element, (b) element response

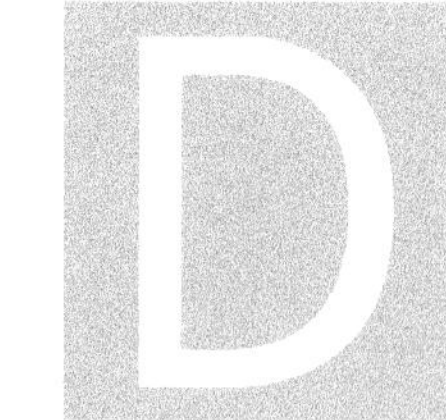

Input parameters used in the strength model

D.1 Parameters used for the strength model with dataset B

The following parameters for the distributions of the different variables were employed for the simulations of GLT beams with wood material corresponding to dataset B. The parameters of the distributions in Tables D.1 and D.2 correspond to parameters of the *standardized* form of the distribution, as defined in the SciPy software package (Virtanen et al., 2020).

Table D.1. Parameters used for the LS13 boards

Variable	Distribution	location	scale	shape	shape 2
Length	Lognorm	9.31×10^{1}	3.73×10^{2}	3.98×10^{-1}	–
$E_{t,0}$	Lognorm	6.22×10^{3}	6.58×10^{3}	4.26×10^{-1}	–
$f_{t,0}$	Lognorm	-3.52×10^{1}	8.89×10^{1}	1.84×10^{-1}	–
$f_{c,0}$	Lognorm	2.61×10^{1}	2.39×10^{1}	9.70×10^{-2}	–
$f_{t,\mathrm{fj}}$	Lognorm	2.47×10^{1}	1.52×10^{1}	5.57×10^{-1}	–
$f_{t,0,\mathrm{cell}}$	Beta	2.01×10^{1}	9.99×10^{1}	1.85	5.91×10^{-1}
$f_{c,0,\mathrm{cell}}$	Weibull	4.23×10^{1}	1.26×10^{1}	3.32	–
$\widetilde{E}_{t,0,\mathrm{cell}}$	Log-Gamma	9.63×10^{-1}	6.18×10^{-2}	7.90×10^{-1}	–

Table D.2. Parameters used for the LS10 boards

Variable	Distribution	location	scale	shape	shape 2
Length	Lognorm	9.31×10^{1}	3.73×10^{2}	3.98×10^{-1}	–
$E_{t,0}$	Lognorm	-3.63×10^{4}	4.90×10^{4}	4.07×10^{-2}	–
$f_{t,0}$	Weibull	2.13×10^{1}	2.66×10^{1}	1.70	–
$f_{c,0}$	Lognorm	2.61×10^{1}	2.39×10^{1}	9.70×10^{-2}	–
$f_{t,\text{fj}}$	Lognorm	1.08×10^{1}	3.24×10^{1}	2.96×10^{-1}	–
$f_{t,0,\text{cell}}$	beta	2.13×10^{1}	9.87×10^{1}	1.55	7.94×10^{-1}
$f_{c,0,\text{cell}}$	Weibull	4.23×10^{1}	1.26×10^{1}	3.32	–
$\widetilde{E}_{t,0,\text{cell}}$	Log-Gamma	9.63×10^{-1}	6.18×10^{-2}	7.90×10^{-1}	–

D.2 Parameters used for the strength model with dataset C

The parameters shown in Table D.3 correspond to the distributions of the different variables that were employed for the simulations of GLT beams with wood material corresponding to dataset C.

Table D.3. Parameters used for the boards of dataset C

Variable	Distribution	location	scale	shape	shape 2
Length	Lognorm	4.15×10^{2}	1.36×10^{3}	2.83×10^{-1}	–
$E_{t,0}$	Lognorm	-2.24	1.13×10^{4}	1.82×10^{-1}	–
$f_{t,0}$	Weibull	8.33	2.45×10^{1}	2.18	–
$f_{c,0}$	Weibull	3.45×10^{1}	1.28×10^{1}	5.22	–
$f_{t,\text{fj}}$	Weibull	7.66	3.82×10^{1}	3.56	–
$f_{t,0,\text{cell}}$	Beta	1.06×10^{1}	7.92×10^{1}	1.97	1.00
$f_{c,0,\text{cell}}$	Weibull	3.45×10^{1}	1.94×10^{1}	5.22	–
$\widetilde{E}_{t,0,\text{cell}}$	Log-Gamma	9.63×10^{-1}	6.18×10^{-2}	7.90×10^{-1}	–

APPENDIX

Reordering of vectors to maintain global correlation

In Section 6.4.4 an algorithm is described to reorder a given vector $\underline{\boldsymbol{A}}$ so that it will match another, equally long vector $\underline{\boldsymbol{B}}$ (assuming both are generated by the same statistical distribution). In the used case, the vector $\underline{\boldsymbol{A}}$ corresponds to the tensile strength of each simulated board (i.e. the minimum value of each simulated $f_{t,0}$ profile), whilst the vector $\underline{\boldsymbol{B}}$ contains the values for $f_{t,0,\mathrm{glob}}$, which were simulated to be correlated to the global MOE, $E_{t,0,\mathrm{glob}}$. If we can find the indices that reorder $\underline{\boldsymbol{A}}$ into $\underline{\boldsymbol{A}}' \approx \underline{\boldsymbol{B}}$, then we can use those indices to reorder the generated $f_{t,0,\mathrm{cell}}$ profiles, too. The steps to find said indices $\underline{\boldsymbol{q}}$ are:

(i) $\underline{\boldsymbol{i}} \longleftarrow \mathrm{argsort}(\underline{\boldsymbol{B}})$

(ii) $\underline{\boldsymbol{j}} \longleftarrow \mathrm{argsort}(\underline{\boldsymbol{A}})$

(iii) $\underline{\boldsymbol{r}} \longleftarrow \mathrm{argsort}(\underline{\boldsymbol{i}})$

(iv) $\underline{\boldsymbol{q}} \longleftarrow \underline{\boldsymbol{j}}[\,\underline{\boldsymbol{r}}\,]$

(v) $\underline{\boldsymbol{A}}' \longleftarrow \underline{\boldsymbol{A}}[\,\underline{\boldsymbol{q}}\,]$

The function `argsort()` returns the indices that sort a given vector, whilst the notation $\underline{\boldsymbol{X}}[\,\underline{\boldsymbol{q}}\,]$ denotes that a vector $\underline{\boldsymbol{X}}$ is reordered according to the indices contained in $\underline{\boldsymbol{q}}$. These five steps are explained in Fig. E.1 in a visual manner.

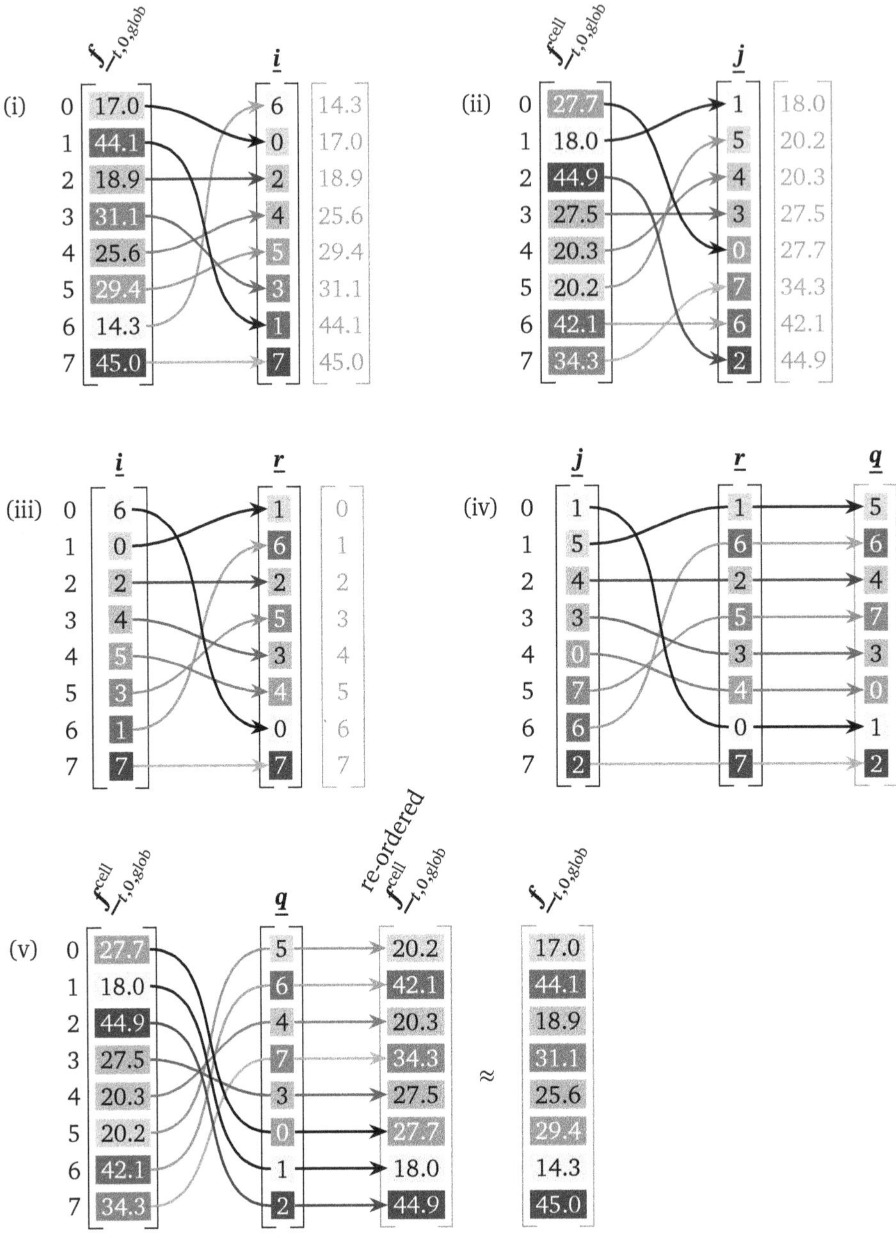

Figure E.1. Steps (i)–(v) from Section 6.4.4 applied to a small generated dataset and visual explanation of each step

www.ingramcontent.com/pod-product-compliance
Ingram Content Group UK Ltd.
Pitfield, Milton Keynes, MK11 3LW, UK
UKHW061827190726
13853UKWH00009B/2477

9 783736 97588(